D

OF INTERN

AND

STAT

STATISTIC

NUMBER 12

World Statistics in Brief

United Nations
statistical pocketbook

TWELFTH EDITION

United Nations

New York - 1988

Note

Symbols of United Nations documents are composed of capital letters combined with figures. Mention of such a symbol indicates a reference to a United Nations document.

General disclaimer

The designations employed and the presentation of material in this publication do not imply the expression of any opinion whatsoever on the part of the Secretariat of the United Nations concerning the legal status of any country, territory, city or area or of its authorities, or concerning the delimitation of its frontiers or boundaries.

ST/ESA/STAT/SER.V/12

UNITED NATIONS PUBLICATION
Sales No. E.88.XVII.8

00500

Inquiries should be directed to:

PUBLISHING SERVICE
UNITED NATIONS
NEW YORK, N.Y. 10017

ISBN 92-1-161294-2

Preface

This publication is the twelfth in a series of annual compilations of basic international statistics issued under the title *World Statistics in Brief (United Nations Statistical Pocketbook).* The compilation was undertaken in response to General Assembly resolution 2626 (XXV), in which the Secretary-General was requested, inter alia, to supply adequate basic national data that would increase international public awareness of countries' development efforts. It is therefore hoped that the *World Statistics in Brief* will serve as an appropriate vehicle for disseminating, in compact form, important basic facts relating to various countries.

The data shown herein were selected from the wealth of international statistical information compiled regularly by the Statistical Office of the United Nations, the statistical services of the specialized agencies and other international organizations. The white pages of the Pocketbook show important and frequently consulted statistical indicators for 167 countries. The insert contains demographic, economic and social statistics for the world as a whole and selected regions of the world.

This issue of the Pocketbook generally covers the years 1976, 1981 and 1986 (or, in some cases, 1985) for the countries, for the world and for regions. The statistics included for each year are those most recently compiled and published by the United Nations and the specialized agencies. The aim is to present, over the period covered, time series that are as nearly comparable internationally as the available statistics permit.

Readers wishing to have more detailed statistics and descriptions are referred to the other, more specialized publications listed on page ii.

It is hoped that the public in general and students in particular will find this dissemination of basic statistics on countries informative, useful and rewarding.

Introduction

Although considerable progress has been made towards standardization of definitions, those employed at the national level in the collection of data often differ significantly from country to country. In this Pocketbook no attempt is made to indicate where deviations occur from the standard definitions. Short descriptions of the standard definitions for selected items are provided in the "General Notes". Additional information can also be found in the *1977 Supplement to the Statistical Yearbook* and the *Monthly Bulletin of Statistics*, published by the United Nations. Various regional groupings of countries are used in one part of the *Pocketbook* and brief notes on the groupings are also provided in the "General Notes" (P. 86).

Readers interested in more detailed figures than those presented in this volume and information regarding their coverage may find it useful to consult the following major publications:

United Nations:
Statistical Yearbook
Demographic Yearbook
Industrial Statistics Yearbook
National Accounts Statistics (series)
International Trade Statistics Yearbook
Energy Statistics Yearbook
Monthly Bulletin of Statistics
Population and Vital Statistics Report

International Labour Office:
Year Book of Labour Statistics

Food and Agriculture Organization of the United Nations:
Production Yearbook
Yearbook of Fishery Statistics -- Catches and Landings
Yearbook of Forest Products
Fertilizer Yearbook
Monthly Bulletin of Statistics

United Nations Educational, Scientific and Cultural Organization:
Statistical Yearbook

International Civil Aviation Organization:
Civil Aviation Statistics of the World: Statistical Yearbook

World Health Organization:
World Health Statistics Annual
World Health Statistics Quarterly

International Monetary Fund:
International Financial Statistics
Direction of Trade Statistics

World Bank:
World Tables – Volume II, Social Data

International Sugar Organization:
Sugar Yearbook

Contents

	Page
Preface	i
Introduction	ii
Afghanistan	2
Albania	2
Algeria	3
Angola	3
Antigua and Barbuda	4
Argentina	4
Australia	5
Austria	5
Bahamas	6
Bahrain	6
Bangladesh	7
Barbados	7
Belgium	8
Belize	8
Benin	9
Bhutan	9
Bolivia	10
Botswana	10
Brazil	11
Brunei Darussalam	11
Bulgaria	12
Burkina Faso	12
Burma	13
Burundi	13
Cameroon	14
Canada	14
Cape Verde	15
Central African Republic	15
Chad	16
Chile	16
China	17
Colombia	17

Contents (continued)

Page

Comoros 18
Congo 18
Costa Rica 19
Côte d'Ivoire 19
Cuba 20
Cyprus 20
Czechoslovakia 21
Democratic Kampuchea 21
Democratic People's Republic of Korea 22
Democratic Yemen 22
Denmark 23
Djibouti 23
Dominica 24
Dominican Republic 24
Ecuador 25
Egypt 25
El Salvador 26
Equatorial Guinea 26
Ethiopia 27
Fiji 27
Finland 28
France 28
Gabon 29
Gambia 29
German Democratic Republic 30
Germany, Federal Republic of 30
Ghana 31
Greece 31
Grenada 32
Guatemala 32
Guinea 33
Guinea-Bissau 33
Guyana 34
Haiti 34
Honduras 35
Hungary 35

Contents (continued)

	Page
Iceland	36
India	36
Indonesia	37
Iran (Islamic Republic of)	37
Iraq	38
Ireland	38
Israel	39
Italy	39
Jamaica	40
Japan	40
Jordan	41
Kenya	41
Kiribati	42
Kuwait	42
Lao People's Democratic Republic	43
Lebanon	43
Lesotho	44
Liberia	44
Libyan Arab Jamahiriya	45
Liechtenstein	45
Luxembourg	46
Madagascar	46
Malawi	47
Malaysia	47
Maldives	48
Mali	48
Malta	49
Mauritania	49
Mauritius	50
Mexico	50
Mongolia	51
Morocco	51
Mozambique	52
Nauru	52
Nepal	53
Netherlands	53

Contents (continued)

	Page
New Zealand	54
Nicaragua	54
Niger	55
Nigeria	55
Norway	56
Oman	56
Pakistan	57
Panama	57
Papua New Guinea	58
Paraguay	58
Peru	59
Philippines	59
Poland	60
Portugal	60
Qatar	61
Republic of Korea	61
Romania	62
Rwanda	62
Saint Kitts and Nevis	63
Saint Lucia	63
Saint Vincent and the Grenadines	64
Samoa	64
Sao Tome and Principe	65
Saudi Arabia	65
Senegal	66
Seychelles	66
Sierra Leone	67
Singapore	67
Solomon Islands	68
Somalia	68
South Africa	69
Spain	69
Sri Lanka	70
Sudan	70
Suriname	71
Swaziland	71

Contents (continued)

	Page
Sweden	72
Switzerland	72
Syrian Arab Republic	73
Thailand	73
Togo	74
Tonga	74
Trinidad and Tobago	75
Tunisia	75
Turkey	76
Tuvalu	76
Uganda	77
USSR	77
Byelorussian Soviet Socialist Republic	78
Ukrainian Soviet Socialist Republic	78
United Arab Emirates	79
United Kingdom	79
United Republic of Tanzania	80
United States	80
Uruguay	81
Vanuatu	81
Venezuela	82
Viet Nam	82
Yemen	83
Yugoslavia	83
Zaire	84
Zambia	84
Zimbabwe	85
General notes	86
Technical notes	87
Conversion coefficients and factors	94
Symbols, conventions and abbreviations	94
Selected list of recurrent publications	95

Data on world and regions

WORLD

Selected Series	Units	1976	1981	1986
AREA				
Surface area	Million km2	135.8	135.8	135.8
Agricultural area	Million HA	4546	4629	4647 a
Forests, wood	Million HA	4145	4097	4086 a
DEMOGRAPHY				
Population				
Total	Millions	4076	4508	4917
Density	Per km2	31	33	36
Econ. active	Millions	1669	1855	2202
Urban (d)	Per cent	38	40	41
Rates: (b)				
Population growth	Per cent pa	2.0	1.8	1.7
Crude birth	Per 1 000	32.7	28.9	27.3
Crude death	Per 1 000	12.7	11.4	10.6
Life expectancy at birth (b)	Years:Males/ Females	55/57	57/59	58/60
PRODUCTION				
Index numbers:				
Total agriculture	1979-81 = 100	91	103	115
Total food	1979-81 = 100	91	102	116
Per caput agriculture	1979-81 = 100	98	101	105
Per caput food	1979-81 = 100	98	101	105
Mining, quarrying (c)	1980 = 100	99.0	92.1	89.9
Manufacturing (c)	1980 = 100	86.0	100.5	116.6
Agricultural:				
Cereals, total	Mill MT	1469.88	1653.61	1867.00
Wheat	Mill MT	418.00	453.82	535.84
Rice, paddy	Mill MT	350.17	411.81	475.53
Barley	Mill MT	184.43	155.16	180.44
Maize	Mill MT	333.08	452.08	480.61
Oats	Mill MT	49.61	42.94	47.77
Sorghum	Mill MT	51.89	72.23	71.44
Roots and tubers, tot.	Mill MT	563.26	553.99	592.42
Potatoes	Mill MT	291.42	287.82	308.55
Pulses, total	Mill MT	49.73	42.32	55.20
Soybean	Mill MT	63.02	88.48	95.52
Vegetables, melons, tot.	Mill MT	312.29	358.66	414.10
Fruits (exc. melons), tot.	Mill MT	259.70	284.49	325.81
Citrus fruits, total	Mill MT	48.69	56.11	59.94
Vegetable fibres, total	Mill MT	14.66	17.30	16.95
Cotton, lint	Mill MT	12.26	15.32	15.05
Coffee, green	1 000 MT	3659	6032	5188
Tea	1 000 MT	1633	1864	2296
Tobacco leaves	1 000 MT	5680	5998	6109
Milk, total	Mill MT	437.09	471.69	520.86
Hen eggs	1 000 MT	23922	27612	31385
Wool, greasy	1 000 MT	2607	2820	3004
Natural rubber	1 000 MT	3596	3753	4372
Round wood	Mill Cu M	2706	2945	3165 a
Livestock numbers and fish catches:				
Cattle	Millions	1209	1214	1272
Pigs	Millions	638	778	822
Sheep	Millions	1034	1122	1146
Goats	Millions	403	471	492
Fish catches	Mill MT	69.59	74.84	91.46

a: 1985; **b:** data refer to 1970-75, 1975-80, 1980-85; **c:** Excluding Albania, China, Democratic Pe[ople's] Republic of Korea and Viet Nam; **d:** Data refer to 1975, 1980 and 1985 respectively.

WORLD (continued)

Selected Series	Units	1976	1981	1986
PRODUCTION (cont'd)				
Mining and quarrying:				
Hard coal	Mill MT	2420.78	2727.75	3194.52
Crude petroleum	Mill MT	2870.19	2795.03	2800.08
Natural gas	1000 TJ	48043	54453	61625
Iron ore	Mill MT	535.66	534.62	524.14a
Manufacturing:				
Meat	Mill MT	122.80	135.02	154.97
Wheat flour	Mill MT	166.55	193.18	212.53a
Sugar, raw	Mill MT	82.40	92.77	100.22
Cement	Mill MT	748.46	877.44	962.30a
Pig iron ferro-alloys	Mill MT	504.39	519.62	505.99a
Crude steel	Mill MT	672.95	696.81	679.48a
Passenger cars	Thousands	28949	27831	31937a
Commercial vehicles	Thousands	8841	9638	12228a
Total primary energy	Mill MT oil equivalent	5896	6233	6887a
Electricity	Billion kWh	6983	8384	9962a
EXTERNAL TRADE AND TRANSPORT				
Imports (c.i.f.)	Bill US$	1014	2038	2216
Exports (f.o.b.)	Bill US$	990	1972	2111
Railways:				
Freight net ton-km.	Billions	6120	6810	7285a
Sea-borne shipping:				
Loaded: Dry cargo	Mill MT	1522	1823	1937a
Tankers	Mill MT	1843	1680	1424a
Unloaded: Dry cargo	Mill MT	1518	1867	1991a
Tankers	Mill MT	1834	1692	1436a
Civil aviation:				
Kilometres flown	Millions	7520k	9030	10417a
Passenger-km	Billions	575k	945	1167a
Motor vehicles in use:(g)				
Passenger cars	Millions	272.2	335.1	373.5a
Commercial vehicles	Millions	72.0	97.0	114.5a
CONSUMPTION (Per caput)				
Sugar	Kilograms	20.4	19.8	20.4
Fertilizers	Kilograms	22k	25	27a
Steel	Kilograms	157.22k	159.75p	100.44b
Newsprint	Kilograms	5.3k	5.8p	6.1b
Commercial energy	Kilograms	1327	1284	1327
Food supply: Calories (c)				
Vegetable products	Numbers	2111d	2253e	2258q
Animal products	Numbers	395d	412e	407q
EDUCATION				
Students enrolled: (f,g)				
First level	Per cent: M/F	103/86k	102/86p	106/91a
Second level	Per cent: M/F	48/38k	51/40p	51/41a
Third level	Per cent: M/F	12/8k	13/10p	13/10a
Public exp. on educ. (h)	As % of GNP	5.6	5.6p	5.8a
COMMUNICATION				
Radios in use (f)	Per 1 000 pop.	238k	277p	330a
TVs in use (f)	Per 1 000 pop.	98k	123p	137a

a: 1985; **b:** 1984; **c:** Daily supplies; **d:** 1974-76; **e:** 1981-83; **f:** Excluding China; **g:** Excluding Democratic People's Republic of Korea; **h:** Excluding Dem. Kampuchea; Democratic People's Republic of Korea; Lao People's Democratic Republic, Lebanon, Mongolia, Mozambique; Viet Nam and South Africa; **k:** 1975; **p:** 1980; **q:** 1983-85

AFRICA

Selected Series	Units	1976	1981	1986
AREA				
Surface area	Million km2	30.3	30.3	30.3
Agricultural area	Million HA	1009	964	973a
Forests, wood	Million HA	639	694	698a
DEMOGRAPHY				
Population:				
Total	Millions	413	484	572
Density	Per km2	14	16	19
Econ. active	Millions	154	179	219
Urban (d)	Per cent	24	27	30
Rates: (b)				
Population growth	Per cent pa	2.7	3.0	3.0
Crude birth	Per 1 000	47.0	46.9	46.4
Crude death	Per 1 000	19.7	18.0	16.5
Life expectancy at birth (b)	Years:Males/ Females	44/47	46/49	48/51
PRODUCTION				
Index numbers:				
Total agriculture	1979-81 = 100	93	103	115
Total food	1979-81 = 100	93	103	115
Per caput agriculture	1979-81 = 100	105	100	97
Per caput food	1979-81 = 100	105	100	97
Mining, quarrying	1980 = 100	-	-	-
Manufacturing	1980 = 100	-	-	-
Agricultural:				
Cereals, total	Mill MT	68.00	78.97	86.00
Wheat	Mill MT	10.46	9.08	11.55
Rice, paddy	Mill MT	7.88	8.44	9.85
Barley	Mill MT	5.07	3.51	6.29
Maize	Mill MT	23.73	34.06	30.83
Oats	Mill MT	.24	.26	.26
Sorghum	Mill MT	9.45	11.96	14.37
Roots and tubers, tot.	Mill MT	78.16	87.26	105.78
Potatoes	Mill MT	4.18	4.99	6.43
Pulses, total	Mill MT	5.28	5.35	7.03
Soybean	Mill MT	.11	.34	.41
Vegetables, melons, tot.	Mill MT	20.05	24.45	28.07
Fruits (exc. melons), tot.	Mill MT	30.38	32.08	41.01
Citrus fruits, total	Mill MT	5.25	4.49	5.20
Vegetable fibres, total	Mill MT	1.37	1.40	1.46
Cotton, lint	Mill MT	1.09	1.16	1.29
Coffee, green	1 000 MT	1260	1269	1360
Tea	1 000 MT	156	198	267
Tobacco leaves	1 000 MT	277	259	339
Milk, total	Mill MT	12.18	13.41	15.80
Hen eggs	1 000 MT	745	969	1271
Wool, greasy	1 000 MT	197	200	223
Natural rubber	1 000 MT	223	194	249
Round wood	Mill Cu M	356	416	459a
Livestock numbers and fish catches:				
Cattle	Millions	159	170	177
Pigs	Millions	8	11	12
Sheep	Millions	158	183	192
Goats	Millions	127	150	161
Fish catches	Mill MT	4.21	4.30	4.38

a: 1985; **b:** data refer to 1970-75, 1975-80, 1980-85 respectively; **d:** Data refer to 1975, 1980 an 1985 respectively.

Selected Series	Units	1976	1981	1986
PRODUCTION (cont'd)				
Mining and quarrying:				
ard coal	Mill MT	82.87	137.53	181.42
rude petroleum	Mill MT	288.38	226.35	235.41
atural gas	1000 TJ	527	918	1818
on ore	Mill MT	35.93	38.97	33.72[a]
Manufacturing:				
eat	Mill MT	5.33[d]	6.80	7.82
heat flour	Mill MT	7.61	10.05	10.40[a]
ugar, raw	Mill MT	5.53	6.45	7.41
ement	Mill MT	22.94	34.67	38.91[a]
ig iron and ferro-alloys	Mill MT	8.24	10.61	11.27[a]
rude steel	Mill MT	8.66	11.27	10.70[a]
assenger cars	Thousands	-	-	-
ommercial vehicles	Thousands	-	-	-
otal primary energy	Mill MT oil equivalent	355	339	404
lectricity	Billion kWh	143	200	234
EXTERNAL TRADE AND TRANSPORT				
nports (c.i.f.)	Bill US$	45	107	74
xports (f.o.b.)	Bill US$	48	85	59
Railways:				
reight net ton-km.	Billions	108	138	126[a]
Sea-borne shipping:				
Loaded: Dry cargo	Mill MT	119	156	164[a]
Tankers	Mill MT	267	279	272[a]
nloaded: Dry cargo	Mill MT	72	112	104[a]
Tankers	Mill MT	43	129	93[a]
Civil aviation:				
ilometres flown	Millions	269[g]	366	400[a]
assenger-km	Billions	18[g]	32	38[a]
Motor vehicles in use:				
assenger cars	Millions	5.1	6.0	6.9[a]
ommercial vehicles	Millions	2.3	2.8	4.0[a]
CONSUMPTION (Per caput)				
ugar	Kilograms	12.9	15.7	14.8
ertilizers	Kilograms	6[g]	7	7[a]
teel	Kilograms	36.64	37.40[p]	18.75[a]
ewsprint	Kilograms	0.8	0.7	0.7[b]
ommercial energy	Kilograms	238	289	297
Food supply: Calories (c)				
egetable products	Numbers	2077[d]	2253[e]	2104[q]
nimal products	Numbers	166[d]	182[e]	174[q]
EDUCATION				
tudents enrolled:				
First level	Per cent: M/F	75/55[g]	90/69[p]	93/76[a]
Second level	Per cent: M/F	20/11[g]	29/17[p]	40/26[a]
Third level	Per cent: M/F	4/1[g]	5/2	6/2[a]
ublic exp. on educ. (f)	As % of GNP	4.7[g]	5.1[p]	5.8[a]
COMMUNICATION				
adios in use	Per 1 000 pop.	69[g]	103[p]	139[a]
Vs in use	Per 1 000 pop.	6.2[g]	17[p]	25[a]

: 1985; **b:** 1984; **c:** Daily supplies; **d:** 1974-76; **e:** 1981-83; **f:** Excluding Mozambique and South frica; **g:** 1975; **p:** 1980; **q:** 1983-85.

AMERICA, NORTH

Selected Series	Units	1976	1981	1986
AREA				
Surface area	Million km^2	24.2	24.2	24.2
Agricultural area	Million HA	618	629	642[a]
Forests, wood	Million HA	718	682	659[a]
DEMOGRAPHY				
Population:				
Total	Millions	347	376	406
Density	Per km2	14	15	17
Econ. active	Millions	138	154	179
Urban (b, d)	Per cent	74	74	74
Rates: (b,c)				
Population growth	Per cent pa	1.1	1.1	0.9
Crude birth	Per 1 000	16.5	16.3	16.0
Crude death	Per 1 000	9.3	9.1	9.1
Life expectancy at birth (b,c)	Years:Males/ Females	68/75	69/77	71/78
PRODUCTION				
Index numbers:				
Total agriculture	1979-81 = 100	90	105	104
Total food	1979-81 = 100	90	105	106
Per caput agriculture	1979-81 = 100	96	104	96
Per caput food	1979-81 = 100	96	104	98
Mining, quarrying (b)	1980 = 100	87.0	105.3	91.1
Manufacturing (b)	1980 = 100	91.0	102.3	115.4
Agricultural:				
Cereals, total	Mill MT	321.97	415.11	404.73
Wheat	Mill MT	85.30	104.22	93.47
Rice, paddy	Mill MT	7.05	10.60	8.26
Barley	Mill MT	19.17	24.72	28.73
Maize	Mill MT	172.91	232.44	231.23
Oats	Mill MT	12.81	10.66	9.60
Sorghum	Mill MT	22.06	29.17	30.55
Roots and tubers, tot.	Mill MT	22.03	22.18	23.44
Potatoes	Mill MT	19.54	19.31	20.36
Pulses, total	Mill MT	2.45	4.10	3.69
Soybean	Mill MT	35.60	55.76	56.23
Vegetables, melons, tot.	Mill MT	29.03	32.39	35.38
Fruits (exc. melons), tot.	Mill MT	39.89	44.76	42.58
Citrus fruits, total	Mill MT	15.03	17.90	13.74
Vegetable fibres, total	Mill MT	2.88	4.06	2.53
Cotton, lint	Mill MT	2.80	4.00	2.40
Coffee, green	1 000 MT	863	933	974
Tea	1 000 MT	-	-	-
Tobacco leaves	1 000 MT	1236	1225	765
Milk, total	Mill MT	70.76	78.86	85.74
Hen eggs	1 000 MT	4862	5473	5567
Wool, greasy	1 000 MT	61	60	47
Natural rubber	1 000 MT	-	-	13
Round wood	Mill Cu M	527	616	677[a]
Livestock numbers and fish catches:				
Cattle	Millions	190	183	171
Pigs	Millions	73	99	91
Sheep	Millions	23	21	21
Goats	Millions	12	14	14
Fish catches	Mill MT	5.18	7.27	8.37

a: 1985; **b:** data refer to Northern America; **c:** Data refer to 1970-75, 1975-80, 1980-85 respectively
d: Data refer to 1975, 1980 and 1985 respectively.

AMERICA, NORTH (continued)

Selected Series	Units	1976	1981	1986
PRODUCTION (cont'd)				
Mining and quarrying:				
Hard coal	Mill MT	618.28	730.67	777.92
Crude petroleum	Mill MT	518.37	615.20	636.48
Natural gas	1000 TJ	23954	22963	19720
Iron ore	Mill MT	87.88	83.69	60.46a
Manufacturing:				
Meat	Mill MT	29.94d	30.28	33.62
Wheat flour	Mill MT	16.89	18.30	19.52a
Sugar, raw	Mill MT	18.10	19.59	20.56
Cement	Mill MT	98.06	102.05	108.49a
Pig iron and ferro-alloys	Mill MT	94.11	83.70	59.32a
Crude steel	Mill MT	134.90	132.27	101.09a
Passenger cars	Thousands	9864	7425	9362a
Commercial vehicles (h)	Thousands	2947	2330	4383a
Total primary energy	Mill MT oil equivalent	1813	1759	1794
Electricity	Billion kWh	2500	2871	3202
EXTERNAL TRADE AND TRANSPORT				
Imports (c.i.f.)	Bill US$	194	398	501
Exports (f.o.b.)	Bill US$	172	347	328
Railways:				
Freight net ton-km.	Billions	1352	1642	1605a
Sea-borne shipping:				
Loaded: Dry cargo	Mill MT	410	534	478a
Tankers	Mill MT	64	106	111a
Unloaded: Dry cargo	Mill MT	188	214	235a
Tankers	Mill MT	492	403	264a
Civil aviation:				
Kilometres flown	Millions	4091g	4771	5677a
Passenger-km	Billions	299g	454	585a
Motor vehicles in use:				
Passenger cars	Millions	123.4	135.1	150.8a
Commercial vehicles	Millions	32.0	38.7	45.8a
CONSUMPTION (Per caput)				
Sugar	Kilograms	45.2/41.1	38.9/43.9	30.6/43.8
Fertilizers	Kilograms	64g	63	57a
Steel	Kilograms	401.01g	415.89	321.53a
Newsprint (f)	Kilograms	38.0g	42.7	49.8b
Commercial energy	Kilograms	6064	4955	4667
Food supply: Calories (c)				
Vegetable products	Numbers	2206d	2342e	2382p
Animal products	Numbers	1000d	989e	996p
EDUCATION				
Students enrolled: (f)				
First level	Per cent: M/F	99/98g	100/100k	104/102a
Second level	Per cent: M/F	103/106g	92/94k	98/101a
Third level	Per cent: M/F	57/48g	51/56k	61/65a
Public exp. on educ. (f)	As % of GNP	6.6g	7.1k	6.8a
COMMUNICATION				
Radios in use (f)	Per 1 000 pop.	1797g	1869k	1977a
TVs in use (f)	Per 1 000 pop.	564g	660k	769a

a: 1985; **b:** 1984; **c:** Daily supplies; **d:** 1974-76; **e:** 1981-83; **f:** Data refer to Northern America; **g:** 1975; **h:** Including assembly and factory sales; **k:** 1980; **p:** 1983-85.

AMERICA, SOUTH

Selected Series	Units	1976	1981	1986
AREA				
Surface area	Million km2	17.8	17.8	17.8
Agricultural area	Million HA	545	584	599a
Forests, wood	Million HA	924	945	917a
DEMOGRAPHY				
Population:				
Total	Millions	214	245	274
Density	Per km2	12	14	15
Econ. active	Millions	70	79	96
Urban (b, d)	Per cent	61	65	69
Rates: (b,c)				
Population growth	Per cent pa	2.5	2.4	2.2
Crude birth	Per 1 000	35.4	33.3	31.8
Crude death	Per 1 000	9.8	8.8	8.2
Life expectancy at birth (b,c)	Years:Males/ Females	59/63	60/65	62/67
PRODUCTION				
Index numbers:				
Total agriculture	1979-81 = 100	88	105	112
Total food	1979-81 = 100	90	104	114
Per caput agriculture	1979-81 = 100	96	102	98
Per caput food	1979-81 = 100	99	102	100
Mining, quarrying (b)	1980 = 100	82	101.1	103.3
Manufacturing (b)	1980 = 100	80	96.9	110.6
Agricultural:				
Cereals, total	Mill MT	65.87	73.98	78.18
Wheat	Mill MT	15.91	11.93	16.75
Rice, paddy	Mill MT	13.52	13.23	15.32
Barley	Mill MT	1.33	.72	.82
Maize	Mill MT	27.23	37.94	37.91
Oats	Mill MT	.72	.59	.85
Sorghum	Mill MT	6.42	8.96	6.05
Roots and tubers, tot.	Mill MT	43.39	42.59	43.25
Potatoes	Mill MT	8.71	10.59	10.1
Pulses, total	Mill MT	2.67	3.27	3.15
Soybean	Mill MT	12.34	19.65	21.46
Vegetables, melons, tot.	Mill MT	10.17	11.05	12.67
Fruits (exc. melons), tot.	Mill MT	36.91	40.02	46.33
Citrus fruits, total	Mill MT	10.82	13.37	17.47
Vegetable fibres, total	Mill MT	1.07	1.29	1.52
Cotton, lint	Mill MT	.82	.97	1.19
Coffee, green	1 000 MT	1117	3115	2046
Tea	1 000 MT	44	39	73
Tobacco leaves	1 000 MT	518	508	530
Milk, total	Mill MT	23.57	23.76	27.01
Hen eggs	1 000 MT	1026	1706	1957
Wool, greasy	1 000 MT	300	305	318
Natural rubber	1 000 MT	26	38	54
Round wood	Mill Cu M	231	279	302a
Livestock numbers and fish catches:				
Cattle	Millions	215	215	252
Pigs	Millions	51	51	53
Sheep	Millions	102	107	107
Goats	Millions	18	20	19
Fish catches	Mill MT	7.23	8.52	14.04

a: 1985; **b:** Data refer to Latin America; **c:** Data refer to 1970-75, 1975-80, 1980-85; **d:** Data refe to 1975, 1980 and 1985 respectively.

AMERICA, SOUTH (continued)

Selected Series	Units	1976	1981	1986
PRODUCTION (cont'd)				
Mining and quarrying:				
Hard coal	Mill MT	8.86	12.00	20.30
Crude petroleum	Mill MT	172.73	178.05	186.66
Natural gas	1000 TJ	1006	1348	1845
Iron ore	Mill MT	93.88	102.88	98.73[a]
Manufacturing:				
Meat	Mill MT	10.17[e]	12.01	12.03
Wheat flour	Mill MT	7.24	10.35	10.88[a]
Sugar, raw	Mill MT	11.29	13.69	13.00
Cement	Mill MT	36.18	49.51	41.20[a]
Pig iron and ferro-alloys	Mill MT	11.19	15.62	22.29[a]
Crude steel	Mill MT	12.44	13.91	16.36[a]
Passenger cars	Thousands	668	743	874[a]
Commercial vehicles (h)	Thousands	508	365	218[a]
Total primary energy	Mill MT oil equivalent	217	238	273
Electricity	Billion kWh	182	277	369
EXTERNAL TRADE AND TRANSPORT				
Imports (c.i.f.)	Bill US$	35	69	44
Exports (f.o.b.)	Bill US$	32	68	55
Railways:				
Freight net ton-km.	Billions	43	49	57[a]
Sea-borne shipping:				
Loaded: Dry cargo	Mill MT	159	201	222[a]
Tankers	Mill MT	120	168	88[a]
Unloaded: Dry cargo	Mill MT	45	76	54[a]
Tankers	Mill MT	56	60	36[a]
Civil aviation:				
Kilometres flown	Millions	385[p]	485	507[a]
Passenger-km	Billions	22[p]	39	41[a]
Motor vehicles in use:				
Passenger cars	Millions	10.4	13.0	17.31[a]
Commercial vehicles	Millions	3.5	4.5	6.0[a]
CONSUMPTION (Per caput)				
Sugar	Kilograms	40.6	42.2	41.6
Fertilizers	Kilograms	13[p]	15	17[a]
Steel	Kilograms	95.42	108.88	22.25[a]
Newsprint (c)	Kilograms	2.7[p]	3.5	3.0[b]
Commercial energy	Kilograms	654	684	708
Food supply: Calories (d)				
Vegetable products	Numbers	2078[e]	2122[f]	2169[q]
Animal products	Numbers	464[e]	469[f]	448[q]
EDUCATION				
Students enrolled: (c)				
First level	Per cent: M/F	100/97[p]	107/104[n]	110/106[a]
Second level	Per cent: M/F	36/34[p]	43/44[n]	51/53[a]
Third level	Per cent: M/F	14/10[p]	15/12[n]	16/14[a]
Public exp. on educ. (c)	As % of GNP	3.5[p]	4.0[n]	4.0[a]
COMMUNICATION				
Radios in use (c)	Per 1 000 pop.	251[p]	276[n]	328[a]
TVs in use (c)	Per 1 000 pop.	84[p]	108[n]	138[a]

a: 1985; **b:** 1984; **c:** Data refer to Latin America and the Caribbean; **d:** Daily supplies; **e:** 1974-76; **f:** 1981-83; estimates; **h:** Including assembly; **n:** 1980;**p:** 1975; **q:** 1983-85.

ASIA (a)

Selected Series	Units	1976	1981	1986
AREA				
Surface area	Million km2	27.6	27.6	27.6
Agricultural area	Million HA	1020	1100	1098 b
Forests, wood	Million HA	603	556	562 b
DEMOGRAPHY				
Population:				
Total	Millions	2354	2625	2866
Density	Per km2	85	95	104
Econ. active	Millions	956	1080	1325
Urban (d, h)	Per cent	28-23	28-25	29-27
Rates: (d,e)				
Population growth	Per cent pa	2.3-2.4	1.4-2.3	1.2-2.2
Crude birth	Per 1 000	32.4-40.6	21.6-37.7	19.0-34.0
Crude death	Per 1 000	10.0-16.2	7.5-14.4	6.8-12.0
Life expectancy at birth (d,e)	Years:Males/ Females	63/65-50/50	66/67-52/53	67/70-55/55
PRODUCTION				
Index numbers:				
Total agriculture	1979-81 = 100	87	104	125
Total food	1979-81 = 100	87	103	125
Per caput agriculture	1979-81 = 100	94	102	113
Per caput food	1979-81 = 100	94	102	113
Mining, quarrying (f,g)	1980 = 100	120.0	82.9	72.4
Manufacturing (f,g)	1980 = 100	79.0	102.9	137.1
Agricultural:				
Cereals, total	Mill MT	559.37	660.53	780.00
Wheat	Mill MT	111.38	140.38	188.00
Rice, paddy	Mill MT	317.59	374.57	436.56
Barley	Mill MT	29.79	17.40	18.73
Maize	Mill MT	53.85	84.62	96.96
Oats	Mill MT	2.50	1.09	1.11
Sorghum	Mill MT	11.91	20.17	18.35
Roots and tubers, tot.	Mill MT	219.48	215.17	218.84
Potatoes	Mill MT	60.41	67.92	72.34
Pulses, total	Mill MT	28.17	21.29	24.55
Soybean	Mill MT	14.05	11.63	15.21
Vegetables, melons, tot.	Mill MT	165.21	193.06	223.17
Fruits (exc. melons), tot.	Mill MT	62.47	79.09	95.59
Citrus fruits, total	Mill MT	10.55	31.05	13.99
Vegetable fibres, total	Mill MT	5.76	6.85	7.79
Cotton, lint	Mill MT	4.78	5.98	7.09
Coffee, green	1 000 MT	375	662	753
Tea	1 000 MT	1334	1482	1789
Tobacco leaves	1 000 MT	2496	2969	3280
Milk, total	Mill MT	59.71	72.25	86.24
Hen eggs	1 000 MT	7324	8009	10549
Wool, greasy	1 000 MT	285	442	459
Natural rubber	1 000 MT	3341	3517	4051
Round wood	Mill Cu M	861	906	985 b
Livestock numbers and fish catches:				
Cattle	Millions	356	364	388
Pigs	Millions	289	366	404
Sheep	Millions	278	327	321
Goats	Millions	228	268	276
Fish catches	Mill MT	28.26	32.34	40.22

a: Excluding USSR; **b:** 1985; **c:** 1982; **d:** Data refer to "East Asia" - "South-Asia"; **e:** Data refe to 1970-75, 1975-80, 1980-85; **f:** Data refer to Asian Middle East and East and South-Asia; **g:** Ex cluding China, Democratic People's Republic of Korea and Viet Nam; **h:** Data refer to 1975, 198 and 1985 respectively.

ASIA (b) (continued)

Selected Series	Units	1976	1981	1986
PRODUCTION (cont'd)				
Mining and quarrying:				
Hard coal	Mill MT	664.90	810.77	1103.14
Crude petroleum	Mill MT	1290.07	1002.96	894.38
Natural gas	1000 TJ	2548	3440	4950
Iron ore	Mill MT	79.19	84.69	103.03a
Manufacturing:				
Meat	Mill MT	25.74	27.47	38.31
Wheat flour	Mill MT	55.71	73.61	91.33a
Sugar, raw	Mill MT	18.14	18.22	23.54
Cement	Mill MT	203.82	293.65	405.00a
Pig iron and ferro-alloys	Mill MT	130.94	142.25	156.44a
Crude steel	Mill MT	148.83	163.53	183.16a
Passenger cars	Thousands	5088	7091	8007a
Commercial vehicles	Thousands	2934	4457	5094a
Total primary energy	Mill MT oil equivalent	1731	1575	1699
Electricity	Billion kWh	1088	1386	1867
EXTERNAL TRADE AND TRANSPORT				
Imports (c.i.f.)	Bill US$	184	455	444
Exports (f.o.b.)	Bill US$	236	539	517
Railways:				
Freight net ton-km.	Billions	642	820	1102a
Sea-borne shipping:				
Loaded: Dry cargo	Mill MT	252	254	320a
Tankers	Mill MT	1169	840	623a
Unloaded: Dry cargo	Mill MT	526	630	785a
Tankers	Mill MT	422	406	446a
Civil aviation:				
Kilometres flown	Millions	884r	1328	1561a
Passenger-km	Billions	78r	178	232a
Motor vehicles in use: (h)				
Passenger cars	Millions	24.2	31.0	41.8a
Commercial vehicles	Millions	14.1	17.6	29.9a
CONSUMPTION (Per caput)				
Sugar	Kilograms	8.6	8.6	11.0
Fertilizers	Kilograms	7r	12	14a
Steel	Kilograms	56.53r	60.71	58.04a
Newsprint	Kilograms	1.6r	2.2p	2.1f
Commercial energy	Kilograms	382	413	482
Food supply: Calories (c)				
Vegetable products	Numbers	2025d	2289e	2247q
Animal products	Numbers	147d	183e	190q
EDUCATION				
Students enrolled: (g,h)				
First level	Per cent: M/F	110/87r	104/83p	110/89a
Second level	Per cent: M/F	43/28r	47/31p	46/32a
Third level	Per cent: M/F	6/3r	7/3p	8/4a
Public exp. on educ. (i)	As % of GNP	4.3r	4.5p	4.7a
COMMUNICATION				
Radios in use (g)	Per 1 000 pop.	60r	95p	145a
TVs in use (g)	Per 1 000 pop.	25r	37p	46a

a: 1985; **b:** Excluding USSR; **c:** Daily supplies; **d:** 1974-76; **e:** 1981-83; **f:** 1984; **g:** Excluding China; **h:** Excluding Democratic People's Republic of Korea; **i:** Excluding Democratic Kampuchea, Democratic People's Republic of Korea, Lao People's Democratic Republic, Lebanon, Mongolia and Viet Nam; **p:** 1980; **q:** 1983-85; **r:** 1975.

EUROPE (a)

Selected Series	Units	1976	1981	1986
AREA				
Surface area	Million km2	4.9	4.9	4.9
Agricultural area	Million HA	230	227	224 b
Forests, wood	Million HA	153	155	155 b
DEMOGRAPHY				
Population:				
Total	Millions	474	485	493
Density	Per km2	96	99	101
Econ. active	Millions	212	219	228
Urban (f)	Per cent	69	70	72
Rates: (c)				
Population growth	Per cent pa	0.6	0.4	0.3
Crude birth	Per 1 000	16.1	14.4	14.0
Crude death	Per 1 000	10.4	10.5	10.7
Life expectancy at birth (c)	Years:Males/ Females	68/74	69/76	70/77
PRODUCTION				
Index numbers:				
Total agriculture	1979-81 = 100	91	100	109
Total food	1979-81 = 100	90	100	109
Per caput agriculture	1979-81 = 100	92	99	107
Per caput food	1979-81 = 100	92	99	107
Mining, quarrying (d)	1980 = 100	73.0/90.0	101.4/99.7	117.9/110.3
Manufacturing (d)	1980 = 100	94.0/82.0	97.7/102.0	106.1/125.5
Agricultural:				
Cereals, total	Mill MT	220.68	246.07	291.61
Wheat	Mill MT	85.91	91.52	116.00
Rice, paddy	Mill MT	1.68	1.76	2.22
Barley	Mill MT	56.33	66.09	70.16
Maize	Mill MT	44.85	54.63	70.72
Oats	Mill MT	14.12	14.67	12.94
Sorghum	Mill MT	.49	.66	.43
Roots and tubers, tot.	Mill MT	112.68	111.94	111.04
Potatoes	Mill MT	112.53	111.79	111.92
Pulses, total	Mill MT	2.30	2.46	5.29
Soybean	Mill MT	.41	.53	1.61
Vegetables, melons, tot.	Mill MT	58.45	64.96	69.87
Fruits (exc. melons), tot.	Mill MT	71.57	67.57	78.91
Citrus fruits, total	Mill MT	6.72	6.73	8.57
Vegetable fibres, total	Mill MT	.40	.38	.49
Cotton, lint	Mill MT	.16	.21	.27
Coffee, green	1 000 MT	-	-	-
Tea	1 000 MT	-	-	-
Tobacco leaves	1 000 MT	832	736	799
Milk, total	Mill MT	168.77	182.70	190.38
Hen eggs	1 000 MT	6663	7270	7379
Wool, greasy	1 000 MT	264	271	304
Natural rubber	1 000 MT	-	-	-
Round wood	Mill Cu M	314	333	349 b
Livestock numbers and fish catches:				
Cattle	Millions	134	133	130
Pigs	Millions	153	174	179
Sheep	Millions	125	137	136
Goats	Millions	11	12	13
Fish catches	Mill MT	13.40	12.53	12.53

a: Excluding USSR; **b:** 1985; **c:** Data relate to 1970-75, 1975-80, 1980-85; **d:** Data refer to "Develope[d] market economies /Centrally planned economies", excluding Albania, Yugoslavia, including th[e] USSR; **f:** Data refer to 1975, 1980 and 1985 respectively.

:UROPE (h)(continued)

;elected Series	Units	1976	1981	1986
PRODUCTION (cont'd)				
Mining and quarrying:				
ard coal	Mill MT	493.39	467.62	463.34
:rude petroleum	Mill MT	82.89	144.88	205.58
atural gas	1000 TJ	8787	9160	8883
on ore	Mill MT	49.21	34.26	29.06a
Manufacturing:				
leat	Mill MT	34.14	39.35	41.61
Vheat flour	Mill MT	35.46	36.56	35.76a
ugar, raw	Mill MT	19.10	24.41	23.08
:ement	Mill MT	257.05	263.84	231.25a
ig iron and ferro-alloys	Mill MT	146.33	142.99	141.24a
rude steel	Mill MT	215.15	210.22	204.97a
assenger cars	Thousands	11721	10931	11997a
ommercial vehicles (g)	Thousands	1620	1562	1613a
otal primary energy	Mill MT oil equivalent	771	875	1002
lectricity	Billion kWh	1913	2195	2529
EXTERNAL TRADE AND TRANSPORT				
nports (c.i.f.)	Bill US$	502	901	1028
xports (f.o.b.)	Bill US$	448	825	1025
Railways:				
reight net ton-km.	Billions	612	618	638a
Sea-borne shipping:				
oaded: Dry cargo	Mill MT	357	426	478a
Tankers	Mill MT	127	164	209a
nloaded: Dry cargo	Mill MT	630	696	721a
Tankers	Mill MT	794	661	574a
Civil aviation:				
ilometres flown	Millions	1616r	1813	1974a
'assenger-km	Billions	135r	210	236a
Motor vehicles in use:				
'assenger cars	Millions	96.9	119.9	137.0a
ommercial vehicles	Millions	12.7	14.4	17.2a
CONSUMPTION (Per caput)				
ugar f	Kilograms	41.5	40.6	41.8
ertilizers	Kilograms	60r	65	65a
teel	Kilograms	389.85r	360.53	343.64a
lewsprint (f)	Kilograms	9.2r	9.5	10.3e
ommercial energy	Kilograms	2973	2937	3108
Food supply: Calories (b)				
egetable products	Numbers	2303c	2297d	2286q
nimal products	Numbers	1059c	1130d	1104q
EDUCATION				
tudents enrolled: (f)				
First level	Per cent: M/F	102/101r	104/103p	104/103a
Second level	Per cent: M/F	78/80r	81/83p	85/87a
Third level	Per cent: M/F	22/19r	23/21p	24/23a
ublic exp. on educ. (f)	As % of GNP	5.8r	5.5p	5.6a
COMMUNICATION				
adios in use (f)	Per 1 000 pop.	380r	466p	564a
Vs in use (f)	Per 1 000 pop.	232r	309p	325a

: 1985; **b:** Daily supplies; **c:** 1974-76; **d:** 1981-83; **e:** 1984; **f:** Including USSR; **g:** Including assembly;
: Excluding USSR; **p:** 1980; **q:** 1983-85; **r:** 1975.

OCEANIA

Selected Series	Units	1976	1981	1986
AREA				
Surface area	Million km2	8.5	8.5	8.5
Agricultural area	Million HA	515	512	503a
Forests, wood	Million HA	144	153	160a
DEMOGRAPHY				
Population:				
Total	Millions	21	23	25
Density	Per km2	2	3	3
Econ. active	Millions	9	7	11
Urban (c)	Per cent	72	71	71
Rates: (b)				
Population growth	Per cent pa	1.9	1.7	1.5
Crude birth	Per 1 000	24.8	21.9	21.1
Crude death	Per 1 000	9.3	9.0	8.4
Life expectancy at birth (b)	Years:Males/ Females	63/68	64/69	65/71
PRODUCTION				
Index numbers:				
Total agriculture	1979-81 = 100	99	101	109
Total food	1979-81 = 100	99	101	107
Per caput agriculture	1979-81 = 100	105	100	100
Per caput food	1979-81 = 100	105	99	98
Mining, quarrying	1980 = 100	99.0	104.4	167.1
Manufacturing	1980 = 100	94.0	100.9	101.4
Agricultural:				
Cereals, total	Mill MT	18.43	24.54	26.49
Wheat	Mill MT	12.14	16.69	17.78
Rice, paddy	Mill MT	.44	.76	.72
Barley	Mill MT	3.19	3.72	4.30
Maize	Mill MT	.41	.33	.46
Oats	Mill MT	1.11	1.66	1.62
Sorghum	Mill MT	1.12	1.21	1.29
Roots and tubers, tot.	Mill MT	2.42	2.71	2.85
Potatoes	Mill MT	.95	1.08	.93
Pulses, total	Mill MT	.21	.23	1.07
Soybean	Mill MT	.04	.07	.11
Vegetables, melons, tot.	Mill MT	1.55	1.74	2.10
Fruits (exc. melons), tot.	Mill MT	3.25	3.72,	4.14
Citrus fruits, total	Mill MT	.44	.52	.68
Vegetable fibres, total	Mill MT	.03	.11	.26
Cotton, lint	Mill MT	.03	1.0	.26
Coffee, green	1 000 MT	44	53	55
Tea	1 000 MT	6	8	10
Tobacco leaves	1 000 MT	19	17	15
Milk, total	Mill MT	13.03	11.87	14.30
Hen eggs	1 000 MT	243	281	237
Wool, greasy	1 000 MT	1066	1082	1188
Natural rubber	1 000 MT	6	5	6
Round wood	Mill Cu M	31	36	37a
Livestock numbers and fish catches:				
Cattle	Millions	44	34	32
Pigs	Millions	4	5	5
Sheep	Millions	205	205	227
Goats	Millions	.1	.4	1
Fish catches	Mill MT	.29	.38	.64

a: 1985; **b:** Data refer to 1970-75, 1975-80, 1980-85; **c:** Data refer to 1975, 1980 and 1985 respectivel

OCEANIA (continued)

Selected Series	Units	1976	1981	1986
PRODUCTION (cont'd)				
Mining and quarrying:				
Hard coal	Mill MT	70.12	87.84	135.51
Crude petroleum	Mill MT	19.45	18.76	26.57
Natural gas	1000 TJ	222	458	711
Iron ore	Mill MT	58.26	59.06	63.50[a]
Manufacturing:				
Meat	Mill MT	4.08	3.88	3.84
Wheat flour	Mill MT	1.46	1.31	1.42[a]
Sugar, raw	Mill MT	4.66	4.00	3.98
Cement	Mill MT	6.16	6.55	6.67[a]
Pig iron and ferro-alloys	Mill MT	7.59	7.52	5.44[a]
Crude steel	Mill MT	8.15	8.19	6.26[a]
Passenger cars	Thousands	369	317	365[a]
Commercial vehicles	Thousands	88	48	17[a]
Total primary energy	Mill MT oil equivalent	74	90	100
Electricity	Billion kWh	102	129	161
EXTERNAL TRADE AND TRANSPORT				
Imports (c.i.f.)	Bill US$	17	34	35
Exports (f.o.b.)	Bill US$	17	28	29
Railways:				
Freight net ton-km.	Billions	34	40	39[a]
Sea-borne shipping:				
Loaded: Dry cargo	Mill MT	172	181	235[a]
Tankers	Mill MT	3	3	7[a]
Unloaded: Dry cargo	Mill MT	22	24	22[a]
Tankers	Mill MT	18	17	14[a]
Civil aviation:				
Kilometres flown	Millions	271[k]	269	296[a]
Passenger-km	Billions	22[k]	31	37
Motor vehicles in use:				
Passenger cars	Millions	6.5	7.3	8.0[a]
Commercial vehicles	Millions	1.7	1.8	2.6[a]
CONSUMPTION (Per caput)				
Sugar	Kilograms	48.7	44.6	44.2
Fertilizers	Kilograms	59[k]	70	65[a]
Steel	Kilograms	343.24[k]	345.00	266.04[a]
Newsprint	Kilograms	30.4[k]	27.0	25.0[a]
Commercial energy	Kilograms	3074	3152	3472
Food supply: Calories (c)				
Vegetable products	Numbers	2015[d]	2152[e]	2143[p]
Animal products	Numbers	1103[d]	1031[e]	990[p]
EDUCATION				
Students enrolled:				
First level	Per cent: M/F	100/96[k]	101/98[h]	99/95[a]
Second level	Per cent: M/F	73/72[k]	70/71[h]	74/75[a]
Third level	Per cent: M/F	23/16[k]	22/19[h]	24/21[a]
Public exp. on educ.	As % of GNP	7.1[k]	6.3[h]	6.3[a]
COMMUNICATION				
Radios in use	Per 1 000 pop.	619[k]	826[h]	1008[a]
TVs in use	Per 1 000 pop.	262[k]	275[h]	333[a]

a: 1985; **b:** 1984; **c:** Daily supplies; **d:** 1974-76; **e:** 1981-83; **h:** 1980; **k:** 1975; **p:** 1983-85.

Data on countries

Afghanistan | Albania

Afghanistan 1976	Afghanistan 1981	Afghanistan 1986	Albania 1976	Albania 1981	Albania 1986
12.08	16.36	18.61	2.45	2.73	3.02
43.8	43.0	41.8	39.9	37.3	35.4
3.9/4.5	4.2/4.7	4.3/4.9	6.4/7.4	6.5/7.5	6.5/7.5
...	54/5[d]	...	50/35[a]	...	...
...	60.1[d]	...	...	...	...
...	12.5[d]	...	...	...	...
13.1	15.6	18.5	32.8	33.4	34.0
0.9	0.6	...	2.4	2.2	2.1
37/37	37/37	39/39	68/72	69/73	70/74
194	194	183	50	45	40
...	...	...	...	...	...
...	...	...	...	...	...
...	...	...	...	...	...
...	...	...	...	...	...
...	...	...	...	...	...
...	...	...	...	...	...
...	...	...	...	...	...
108	101	99	89	99	108
108	101	92	100	97	95
...	...	...	...	...	...
2 347	2 456	2 843	2 657	3 280	4 455
335	622	851	...	...	...
299	694	536	...	...	...
...	108[e]	110[be]	...	...	...
...	...	...	...	...	...
45.0[p]	50.6[p]	50.6[p]	4.10[q]	7.00[q]	7.00[q]
131	274	259	...	...	...
0.93	0.96	0.96	...	...	...
91	9	9[f]	...	...	...
2[a]	4	...	...	...	...
2[a]	2[c]	...	...	...	...
-	3[c]	6[b]	...	...	76[b]
81/96[a]	70/95[d]	...	...	...	...
33/6[a]	46/11	20/9[b]	97/90[k]	97/89[c]	92/86[b]
1.2[a]	1.6	...	...	...	...
17 292[g]	14 471[gh]	...	966[k]	...	...
...	...	...	...	...	...

a: 1975; **b:** 1985; **c:** 1980; **d:** 1979; **e:** Kabul, excluding rent; **f:** 1984; **g:** Personn in government services only; **h:** 1982; **k:** 1977; **p:** Currency: Afghanis; **q:** Currenc Leks.

Algeria

1976	1981	1986
16.52	19.25	22.42[a]
47.6	46.5	45.4
5.7/6.5	5.3/6.2	4.9/5.8
37/4[e]	39/3[f]	34/2[g]
20.3[e]	...	30.2[g]
10.7[e]	...	13.5[g]
40.3	41.2	42.6
3.1	3.1	3.0
53/57	58/62	61/64
112	88	74
15 036[b]	42 113[c]	...
6.1	8.4	...
939[b]	2 256[c]	...
43	40[k]	...
9	6[k]	...
39	41[k]	...
10	12[k]	...
98	100	125
111	97	104
70	108	...
58 215	58 897	82 176
5 341	11 295	10 162
5 335	13 320	7 876
70[e]	114	142
...	...	...
4.36[p]	4.38[p]	4.82[p]
1 765	3 695	1 660
5.47	5.58	5.58
185	951	596
18[b]	43	...
14[b]	33	35[d]
31[b]	52[c]	72[d]
...	43/68[f]	...
70/46[b]	75/54[c]	82/63[d]
4.8[b]	4.1[f]	4.4[d]
4 750[b]	...	...
2 168[b]	2 617	2 710

Angola

1976	1981	1986
6.75[a]	7.94	8.98
43.5	44.2	44.6
4.5/5.4	4.5/5.4	4.5/5.4
50/5[b]	...	...
...	...	...
...	...	...
17.8	21.0	24.5
3.1	3.4	2.5
39/42	40/44	42/46
160	149	137
...	...	...
...	...	...
...	...	...
...	8	8
...	44	50
...	28	22
...	3	3
106	99	102
114	97	87
...	...	...
7 696	7 333	14 158
429[b]	1 678	682[g]
1 012	1 874	1 840[g]
...	...	...
...	...	...
27.47[bq]	...	29.62[q]
...	...	...
...	...	...
...	...	...
...	27	22[h]
4[e]	7	...
-	3.9[c]	4.6[d]
75/88	64/81[c]	...
...	...	...
...	5.4[f]	5.1[h]
...	...	...
...	...	...

: United Nations estimate; **b:** 1975; **c:** 1980; **d:** 1985; **e:** 1977; **f:** 1982; **g:** 1983; **h:** 984; **k:** 1979; **p:** Currency: Algerian dinars; **q:** Currency: Kwanzas.

Antigua and Barbuda

1976	1981	1986
0.07	0.08	0.08
...	...	...
...	...	...
...	...	...
...	...	...
...	...	...
30.8a	30.8	...
0.3	1.1	1.3p
...	...	...
...	13.6	...
51a	107c	161k
0.1	4.7	3.6
729a	1 427c	2 038k
...	45	27k
...	6	5k
...	8	7k
...	5	4k
...	...	...
...	...	...
...	...	...
...	...	...
67a	...	...
28a	...	...
82.8e	160.7e	...
...	...	...
2.372aq	2.700q	2.700q
7a	7	28
...	...	...
...	85	149g
87	...	...
...	...	...
211	213c	238b
...	...	...
...	...	...
4.0a	3.1c	2.9k
2 483h	2 310	1 887k
2 078	2 142	2 105

Argentina

1976	1981	1986
26.48	28.69	31.03
29.2	30.0	31.0
10.6/12.3	10.8/13.1c	11.1/13.8
56/21	53/19c	57/20
...	12.0c	...
...	21.3c	...
80.6	82.7	84.6
1.6	1.6	1.5
65/72	66/73	67/74
41	36	32
35 750a	154 005c	64 829n
3.1	2.1	-1.4
1 372a	5 454c	2 188n
26a	18	18n
9a	9	13n
36a	29	37n
32a	24	31n
94	102	108
100	101	99
102	85	94
28 436	36 042	39 092
3 034	9 430	4 724
3 916	9 143	6 852
30d	2 040d	256 314d
...	100	138 900
0.006ar	0.001r	1.257r
1 445	3 268	2 718
4.00	4.37	4.37
1 250	1 145	1 600
87a	160	168b
78a	98	...
154a	182c	213b
6/7a	6/6c	...
84/86a	...	92/96b
2.4a	3.1c	3.7k
530a	...	...
3 281	3 252	3 195

a: 1975; **b:** 1985; **c:** 1980; **d:** Buenos Aires; metropolitan area; **e:** Index base 1978 = 100; **f:** Domestic agricultural products; including exported products; **g:** By a only; **h:** 1977; **k:** 1984; **n:** 1983; **p:** 1981-83; **q:** Currency: East Caribbean dollars; **r:** Currency: Australes.

ustralia			Austria		
1976	1981	1986	1976	1981	1986
14.03	14.92	15.97	7.57	7.56	7.56
27.6	25.3	23.6	23.3	20.5	18.6
1.2/14.4	12.0/15.4	12.8/16.2	16.6/23.9	15.0/23.0	15.1/24.1
57/32	57/35	75/46[b]	...	57/35	57/33[b]
6.7	5.7	6.0[b]	12.5	8.5	8.8[b]
21.7	19.8	19.7[b]	32.3	32.4	29.9[b]
85.9	85.8	85.5	53.1	54.6	56.1
1.5	1.3	1.3	-0.0	-0.0	-0.0
70/77	72/79	72/79	68/76	70/77	70/78
12	10	8	16	12	10
100 127	156 987	167 441[b]	37 671[a]	76 882[c]	66 053[b]
3.6	2.8	2.8	4.2	3.3	1.6
7 348	10 683	10 666[b]	5 009[a]	10 244[c]	8 805[b]
24	26	23[b]	26	25	22[b]
5	5	4[b]	5	4	3[b]
27	28	27[b]	32	31	31[b]
20	19	17[b]	29	27	28[b]
97	101	112	95	99	104
102	99	100	95	99	95
92[d]	101[d]	102[bd]	85	99	110
70 043	85 711	126 003	6 830	6 219	5 671
11 182[e]	23 768[e]	24 109[e]	11 523	21 048	26 843
13 158	21 767	22 496	8 507	15 845	22 522
68.7	109.7	162.4	83.0	106.8	129.0
...	108[f]	152[f]	...	108[g]	110[g]
0.9205[q]	0.887[q]	1.504[q]	16.770[hr]	15.885[hr]	13.710[hr]
2 870	1 671	7 246	3 560	5 285	6 162
7.36	7.93	7.93	20.88	21.11	21.14
532	937	1 429	11 591	14 241	15 092
350[a]	507	551[b]	227[a]	325	411[b]
382[a]	489[c]	550[b]	281[a]	421	492[b]
334[a]	381[c]	446[b]	...	296[ck]	435[b]
...	...	...	...	...	...
98/98[a]	97/98[c]	100/101[b]	82/84[a]	80/82[c]	84/86[b]
6.1[a]	5.9[c]	6.0[n]	4.5[a]	4.8[c]	5.2[b]
656	524	...	540	436[p]	...
3 305	3 376	3 343	3 317	3 453	3 484

1975; **b:** 1985; **c:** 1980; **d:** Fiscal year beginning 1 July; excluding mining; **e:** Im-
orts f.o.b.; **f:** Industrial products (manufacturing industry only); including production
exported products; **g:** Domestic supply; including exported products; excluding pro-
ucts of mining and quarrying; **h:** Selling rate; **k:** Number of licences issued or sets
eclared; **n:** 1984; **p:** 1982; **q:** Currency: Australian dollars; **r:** Currency: schillings.

Bahamas

1976	1981	1986
0.19	0.21	0.24
...	...	...
...	...	...
47/36	47/36	...
...	5.2	...
...	7.6	...
56.7[a]	56.7[c]	...
2.2	2.0	1.8
...	...	...
...	32[c]	23[f]
754[a]	1 475[c]	2 258[b]
-5.1	6.8	1.3
3 696[a]	7 024[c]	9 817[b]
8	15	16[f]
5	5	5[f]
11	13	12[f]
...	...	...
...	...	...
...	...	...
...	...	...
...	...	...
3 124	7 284	3 288
2 992	6 189	2 702
74.7	111.1	140.6
...	...	...
1.00[h]	1.00[h]	1.00[h]
47	100	232
...	...	...
940	1 031	1 375
182[a]	231	275[d]
275[a]	358	424[b]
...	149[c]	222[b]
...	...	...
...	...	...
5.7[a]	5.1[c]	...
1 170	1 218[c]	...
2 324	2 497	2703

Bahrain

1976	1981	1986
0.26	0.35	0.41
43.0	34.7	33.7
3.5/3.7	3.2/3.8	2.8/3.8
...	62/11	43/10
...	...	...
...	...	...
79.3	80.5	81.7
4.9	4.4	3.7
66/70	67/71	68/73
38	32	27
1 184[a]	15 094[c]	17 204[b]
-1.9	11.0	4.2
4 353	10 902[a]	11 882[c]
42	29	40[d]
2	1	1[d]
41	42	32[d]
13	14	11[d]
...	...	...
...	...	...
...	...	...
4 897	5 326	5 563
1 670	3 954	...
1 386	4 177	2 369
69.1	111.3	...
...	...	...
0.396[k]	0.376[k]	0.376[k]
436	1 544	1 489
0.15	0.15	0.15
...	177	385
90[a]	197	248[b]
101	216	174[b]
110[a]	152[d]	394[b]
...	...	...[b]
82/72[a]	92/79[c]	101/97[b]
...	2.4[c]	3.3[b]
1 876	935[e]	771[b]
...	...	...

a: 1975; **b:** 1985; **c:** 1980; **d:** 1983; **e:** 1982; **f:** 1984; **h:** Currency: Bahamian dollar
k: Currency: Bahraini dinars.

Bangladesh			Barbados		
1976	1981	1986	1976	1981	1986
80.81	90.46	100.62	0.25	0.25	0.25
45.9	46.2	45.7	31.5	29.6	27.1
6.2/5.2	5.5/4.8	5.0/4.5	12.5/14.6	12.2/15.9	11.8/15.7
...	57/13	54/5[c]	48/36[a]	52/40[f]	...
...	77.1	57.7[c]	8.1[a]	8.5[e]	8.0[b]
...	4.7	9.1[c]	14.0[a]	13.6	16.8[b]
9.1[a]	10.4	11.9[b]	38.6	40.1	42.2
2.8	2.8	2.7	0.3	0.3	0.6
46/44[a]	47/46	48/47[b]	69/74	70/75	71/76
140[ar]	137[r]	128[br]	27	14	11
8 941[a]	15 094[f]	17 204[b]	402[a]	947[f]	1 230[b]
4.8	4.1	3.7	1.9	4.1	2.5
117[a]	171[f]	170[b]	1 634[a]	3 458[f]	4 862[b]
13	16	13[b]	26	27	16[b]
51	46	52[b]	9	7	6[b]
8	10	8[b]	11	13	14[b]
8	10	8[b]	10	10	9[b]
87	101	116	84	94	92
97	99	100	85	94	90
79[g]	101[g]	103[bg]	90	97	103
750	1 222	2 920	24	36	103
788	1 813	2 014	236	572	587
432	662	955	87	194	275
62.8[h]	113.2[h]	182.3[h]	65.2	114.6	146.6
...	108[k]	183[k]	...	...	...
14.95[s]	19.85[s]	30.80[s]	2.00[t]	2.01[t]	2.01[t]
289	138	409	28	101	152
...	0.05	0.05	...	0.01	0.01
37	49	129	224	353	370
0.4[a]	0.9	...	95[a]	107	150[b]
1[a]	1	...	163[a]	291	...
0.3[a]	0.9[f]	3[b]	188[a]	201[f]	257[b]
63/87[n]	60/82	...	...	...	...
60/28[a]	50/26[f]	46/29[b]	91/95[a]	94/95	102/101[q]
0.7[a]	1.0[f]	1.8[b]	5.2[a]	5.4[f]	5.4[q]
12 689	8 908	...	1 470	1 167[p]	...
1 843	1 850	1 859	2 947	3 125	3 129

a: 1975; **b:** 1985; **c:** 1983-84; **d:** 1983; **e:** Including mining and quarrying; **f:** 1980; **g:** Fiscal year beginning 1 July; **h:** Dhaka; government officials; **k:** Domestic supply; ncluding exported products; excluding products of mining and quarrying; **n:** 1974; **p:** 1982; **q:** 1984; **r:** United Nations estimate; **s:** Currency: Takas; **t:** Currency: Barbadian dollars.

Belgium 1976	Belgium 1981	Belgium 1986	Belize 1976	Belize 1981	Belize 1986
9.81	9.85	9.91[a]	0.14	0.15	0.17
22.2	20.2[c]	18.9	...	...	...
16.4/21.7	15.4/21.0	16.3/22.0	...	...	...
55/28[b]	54/31[d]	53/23[e]	...	50/15[d]	...
3.4[b]	2.6	2.5[e]	...	31.2[d]	...
30.0[b]	23.2	20.9[e]	...	10.1[d]	...
94.6	95.4	96.3	...	...	...
0.1	0.1	0.1	1.8	2.1	2.5
69/76	70/77	71/78	...	...	...
13	11	10	...	...	...
61 751[b]	117 792[d]	79 076[c]	104[b]	171[d]	195[c]
4.0	2.7	0.7	5.3	4.4	2.2
6 304[b]	11956[d]	7 985[c]	794[b]	1 179[d]	1 196[c]
22	18	16[c]	32	27	17[e]
3	2	2[c]	21	23	20[e]
20	26	26[c]	11	12	11[e]
37	22	23[c]	10	11	8[e]
96[f]	102[f]	104[f]	...	...	...
97[f]	102[f]	104[f]	...	...	...
95	97	105	...	...	...
5 883	5 014	7 040	...	...	...
35 511[f]	62 464[f]	68 667[f]	89[b]	131[g]	...
32 888[f]	55 705[f]	68 876[f]	67[b]	94[g]	...
80.2	107.6	142.3[n]	...	...	...
...	108[h]	132[h]	...	...	...
35.98[q]	38.46[q]	40.41[q]	1.977[br]	2.00[r]	2.197[cr]
3 491	4 952	5 538	...	13[d]	9[e]
42.17	34.18	34.18	...	...	...
7 312	6 550	7 352	...	62	94
267[b]	362	367[c]	46[b]	79	...
285[b]	387	440[c]	40[b]	60	...
260[bk]	298[dk]	300[ck]	...	...	...
...	...	...	...	...	...
93/92[b]	93/95[d]	94/97[c]	...	...	...
5.7	5.9[d]	5.8[c]	...	...	...
530[n]	371[gn]	...	3 150[b]	...	...
3 542[f]	3 583[f]	3 695[f]	2 652	2 711	2 546

a: United Nations estimate; **b:** 1975; **c:** 1985; **d:** 1980; **f:** Including data for Luxem bourg; **g:** 1982; **h:** Domestic supply; **k:** Number of licenses issued or sets declared **n:** Including physicians practicing dentistry; **q:** Currency: Belgian francs; **r:** Currency Belizean dollars.

Benin 1976	Benin 1981	Benin 1986	Bhutan 1976	Bhutan 1981	Bhutan 1986
3.20	3.52	4.04	1.18[a]	1.31[a]	1.45[a]
44.9	45.9	46.8	40.4	40.4	40.0
5.0/5.6	4.6/5.1	4.3/4.8	4.8/5.8	4.9/5.8	5.0/5.9
51/42	44/23[d]	...	59/39	...	...
...	...	...	...	...	...
...	...	...	...	...	...
21.5	28.2	35.2	3.5[b]	3.9	4.5[c]
2.8	3.0	3.1	2.1	2.0	2.0
40/44	42/46	44/48	43/42	45/43	47/45
130	120	110	153[a]	147[a]	139[a]
528[b]	1 162[k]	1 153[e]	...	149	187[c]
4.8	-0.8	0.7	5.2	9.3	5.5
174[b]	333[k]	269[e]	...	113	132
16	17	...	...	...	...
40	39	...	...	...	...
8	7	...	...	...	...
7	...	...	...	...	...
90	97	145	91	102	115
100	94	116	98	100	102
...	...	...	...	...	...
...	223[e]	365	1[b]	1	1
212	542	...	...	...	...
38	34	...	...	...	...
...	...	...	...	...	...
...	...	...	...	...	...
248.50[p]	287.40[p]	322.75[p]	...	8.90[fq]	12.55[fq]
19	58	4	...	...	...
0.00	0.01	0.01	...	...	...
19	46	...	...	2[g]	2[h]
6[b]	9	...	...	...	...
...	...	5[c]	...	...	...
...	1.4[k]	3.7[c]	...	...	...
...	75/90[d]	...	...	...	...
43/19[b]	58/25[k]	61/29[h]	8/3	...	20/10[h]
4.8[b]	5.1[k]	...	...	...	...
31 920[b]	17 500[k]	...	...	...	...
2 017	2 140	2 136	...	...	...

: United Nations estimate; **b:** 1975; **c:** 1985; **d:** 1979; **e:** 1983; **f:** Operational exchange rate for United Nations programmes as of 1 October; **g:** 1982; **h:** 1984; **k:** 1980; **p:** Currency: CFA francs; **q:** Currency: Ngultrum.

Bolivia 1976	Bolivia 1981	Bolivia 1986	Botswana 1976	Botswana 1981	Botswana 1986
5.03	5.76	6.55	0.71	0.94	1.13
43.2	43.5	43.8	50.3	49.5	49.1
5.0/5.6	4.9/5.5	4.8/5.5	2.9/3.7	2.9/3.7	2.8/3.9
51/14	...	48/14[c]	49/46[a]	43/26	38/36[c]
46.2	...	...	...	48.5	...
13.8	...	...	...	5.7	...
41.5	44.3	47.8	12.5	15.3	19.2
2.6	2.7	2.8	3.8	3.8	3.7
47/51	49/53	51/55	51/54	53/56	55/58
138	124	110	82	76	67
2 450[a]	5 018[b]	6 266[c]	369[a]	1 003[b]	1 165[k]
5.9	2.5	-2.8	9.7	12.3	16.4
501[a]	901[b]	984[c]	489[a]	1 096[b]	1 135[k]
20	14	12[c]	25	39	26[k]
18	19	26[c]	24	11	6[k]
27	29	22[c]	24	28	41[k]
14	16	13[c]	8	9	6[k]
99	105	103	124	105	99
110	103	89	145	101	78
...	...	...	...	...	...
3 417	3 423	3 377	[d]	[d]	[d]
594	917	878	[d]	[d]	[d]
573	984	564	[d]	[d]	[d]
50[e]	1 300[e]	6 833 664[e]	63.7	116.3	158.6
...	135[f]	4 341 186[cf]	...	...	...
...	...	1.92[p]	0.870[q]	0.880[q]	1.84[q]
151	100	235	75	253	1 198
0.41	0.83	0.89	...	...	...
185	155	133	115[g]	227[g]	381[g]
6[a]	15	22[c]	6[a]	30	34[c]
...	25	...	10	...	...
9	54[b]	66[c]	...	...	...
24/49	...	...	...	...	...
...	77/64	80/70[h]	47/56[a]	59/71[b]	72/80[c]
3.5[a]	4.2[b]	0.5[h]	...	...	...
...	...	...	9 970	7 378[b]	...
2 015	2 084	2 114	2 117	2 139	2 164

a: 1975; **b:** 1980; **c:** 1985; **d:** Data included with South Africa; **e:** La Paz; **f:** Domest supply; **g:** Including arrivals for employment and transit; **h:** 1984; **k:** 1983; **p:** Curre cy: Bolivianos; **q:** Currency: Pula.

Brazil 1976	Brazil 1981	Brazil 1986	Brunei Darussalam 1976	Brunei Darussalam 1981	Brunei Darussalam 1986
107.54	124.07	138.49	0.16	0.19	0.24[a]
40.1	37.7	36.4	...	...	...
5.6/6.1	5.9/6.4	6.3/6.8	...	...	...
...	54/20[d]	...	...	52/19[a]	...
36.2	29.9[d]	...	...	...	...
16.5	17.2[d]	...	...	...	...
61.8	67.5	72.7	70.0[b]	76.3	...
2.3	2.2	2.1	3.7	3.5	...
60/64	61/66	62/68	...	...	...
79	71	63	...	...	...
123 662[b]	239 766[d]	226 787[c]	1 153[b]	4 848[d]	3 0.32[c]
10.7	6.6	5.5	6.9	10.8	-2.1
1 145[b]	1 977[d]	1 673[c]	7 391[b]	24 735[d]	12 847[c]
0	20	18[c]	...	20	21[q]
0	8	11[q]	1	0.7	0.7[q]
0	28	30[q]	88	78	74[q]
0	28	26[q]	12	9	9[q]
83	107	115	...	...	...
98	102	104	...	...	...
79[e]	90[e]	110[e]	...	...	...
17 829	25 599	51 402	18 441	16 328	17 801
13 726	24 079	15 586	253	599	615[c]
10 128	23 680	22 392	1 296	4 066	2 972[c]
19[f]	100[f]	11 084[f]	84.6[g]	109.1	123.9[c]
...	213[h]	11 783[ch]	...	...	...
...	0.13[kv]	14.89[kv]	2.45[bnx]	2.12[nx]	2.16[nx]
6 488	6 604	5 803	...	...	...
1.33	2.20	2.43	...	...	...
556	1 358	1 934	...	287[c]	398[cp]
46[b]	81	83[c]	140[b]	242	363[q]
29[b]	72	84[c]	59[b]	114	146[c]
...	124[d]	184[c]	90[b]	142[d]	151[c]
22/26	24/27[d]	20/23[c]	...	15/31	...
73/72[b]	82/80[d]	85[qr]	...	...	...
...	...	...	1.7	1.6	1.8[q]
1 510	1 139	...	3 059	1 869[s]	...
2 497	2 623	2 629	2 543	2 774	2 790

: United Nations estimate; **b:** 1975; **c:** 1985; **d:** 1980; **e:** Excluding electricity, gas nd water; **f:** Index base: 1981 = 100; **g:** 1977; **h:** Agricultural products and products f manufacturing; **k:** Selling rate; **n:** Operational exchange rate for United Nations rogrammes as of 1 October; **p:** Including nationals residing abroad; **q:** 1984; **r:** Both exes; **s:** 1982; **t:** 1983; **v:** Currency: cruzados; **x:** Currency: Brunei dollars.

Bulgaria

1976	1981	1986
8.76	8.89	8.96
22.0	22.1	22.3
15.1/17.1	14.5/16.8	15.9/18.6
54/48	...	...
23.6	...	...
34.9	...	...
57.5	62.5	66.5
0.3	0.5	0.4
69/74	69/74	70/75
22	18	15
15 145[kn]	21 933[kn]	25 450[bkn]
7.8[n]	5.3[n]	3.8[n]
1 729[p]	2 467[p]	2 840[bp]
...	...	...
21	19	14
51	48	60
...	...	...
97	102	103
95	103	104
80	105	129
9 683	11 630	14 082
5 626[s]	10 801[s]	13 656[b s]
5 382	10 689	13 348[b]
82.3	100.4	108.4
...	...	...
0.97[q]	0.85[q]	1.23[q]
...	...	...
...	...	...
4 033	6 046	7 657
...	72	...
89[a]	170	...
173[af]	186[ef]	187[bf]
...	...	...
...	95/94[e]	101/101[b]
5.1[a]	5.4[e]	6.3[b]
406[a]	394[h]	...
3 530	3 628	3 626

Burkina Faso

1976	1981	1986
5.74	6.20	6.75
43.8	43.9	43.9
4.2/5.0	4.3/5.1	4.4/5.2
48/2	53/24[e]	...
...	...	...
...	...	...
6.3	7.0	7.9
2.0	2.4	2.6
42/45	44/47	46/49
157	150	139
592[a]	1 218[e]	833[g]
4.9	4.3	0.5
106[a]	198[e]	123[g]
15	19	21[g]
41	43	43[g]
15	12	15[g]
14	11	14[g]
88	104	148
95	102	126
...	...	...
...	...	...
144	338	333[b]
54	75	70[b]
...	100.0[d]	108.7[d]
...	...	...
224.27[r]	287.40[r]	322.75[r]
71	71	234
...	0.01	0.01
30	31	60
1.7[a]	5	...
1[a]	...	2[g]
1.0[a]	1.2[e]	5.2[b]
85/97[a]	...	...
12/7[a]	15/9[e]	24/14[b]
2.3	2.4[e]	2.6[b]
52 820	48 386[e]	...
2 152	2 033	1 961

a: 1975; **b:** 1985; **c:** 1979; **d:** Ouagadougou; **e:** 1980; **f:** Number of licences issued or sets declared; **g:** 1984; **h:** 1982; **k:** Million leva; **n:** Net material product; **p:** In national currency; **q:** Currency: Leva; **r:** Currency: CFA francs.**s: F.O.B.**

Burma 1976	Burma 1981	Burma 1986	Burundi 1976	Burundi 1981	Burundi 1986
30.83	35.09	39.41a	3.82	4.23	4.85
40.7	39.4	37.6	42.5	43.6	44.8
5.6/6.7	5.9/7.0	6.1/7.3	4.8/5.9	4.8/5.9	4.8/5.9
53/30	...	...	56/43b	59/62f	60/62g
69.0	63.6e	...	...	...	...
8.0	8.2e	...	...	...	...
23.9	23.9	23.9	3.3	5.3	8.2
2.0	1.9	1.9	1.8	2.8	2.8
53/57	56/59	58/62	43/47	45/48	47/50
75	70	63	130	124	114
3 682b	5 851d	6 812c	415b	961d	...
2.1	6.2	5.5	0.6	5.0	2.0
121b	174d	183c	111b	234d	...
8	20	17c	9	13	...
47	47	48c	57	57	...
11	11	12c	9	8	...
10	9	10c	9	8	...
82	107	141	91	111	113
90	106	127	103	105	97
...	...	...	...	...	...
1 454	1 907	2 582	...	2	3
194	373	304	58	161	205
193	476	265	61	71	169
104.1k	100.3k	123.6ck	49.2h	112.1h	155.1h
...	...	...	...	...	...
6.732p	7.397p	7.039p	90.000q	90.000q	124.165q
118	229	33	49	61	69
0.20	0.25	0.25	0.00	0.02	0.02
16b	43	47	31	37	66
1.2b	2.1	3c	1.4b	3	3c
1b	...	...	1b	...	...
-	0.0d	0.5c	...	...	0.1c
...	...	...	...	57/74f	...
55/50b	58n	...	...	...	...
1.6b	...	...	2.2b	3.0	2.5c
5 370b	4 940	...	...	...	...
2 134	2 375	2 518	2 275	2 344	2 217

Marked break in series; **b:** 1975; **c:** 1985; **d:** 1980; **e:** 1981-82; **f:** 1982; **g:** 1984; Bujumbura;government officials; including indirect taxes; **k:** Rangoon; **n:** Both sexes; Currency: Kyats; **q:** Currency: Burundi francs.

Cameroon | Canada

Cameroon 1976	Cameroon 1981	Cameroon 1986	Canada 1976	Canada 1981	Canada 1986
7.66	8.97	10.45	22.99	24.34	25.61
41.5	42.4	43.3	26.5	22.5	21.5
5.5/6.8	5.6/6.8	5.6/6.7	11.3/13.6	12.2/15.2	13.1/16.5
42/29	50/30[c]	...	56/34	59/40	...
73.8	73.2[c]	...	6.0[a]	5.0[d]	5.3[e]
4.6	4.7[c]	...	21.7[a]	20.9[d]	20.5[e]
26.9	34.7	42.4	75.6	75.7	75.9
2.6	2.7	2.8	1.2	1.1	1.0
47/50	49/53	51/55	71/78	71/79	73/80
111	103	94	12	9	8
3 066[a]	8 502[d]	8 385[n]	167 263[a]	263 242[d]	348 291[b]
3.5	7.7	8.5	5.5	4.0	2.6
404[a]	986[d]	897[n]	7 360[a]	10 927[d]	13 698[b]
21	23	25[n]	24	24	20[b]
33	27	22[n]	4	4	4[e]
11	24	29[n]	24	25	27[e]
9	11	11[n]	18	17	18[e]
96	102	112	99	108	123
110	100	95	105	107	116
...	...	...	91	101	119
108	4 493	9 175	174 992	187 691	220 556
609[f]	1 428[f]	1 705[f]	37 991[h]	66 303[h]	81 099[h]
511[f]	1 122[f]	784[f]	38 957	70 018	86 725
...	1 460[n]	180.5[k]	70.8	112.5	148.9
...	...	...	...	121[p]	129[b]
248.50[s]	287.400[s]	322.750[s]	1.009[t]	1.186[t]	1.380[t]
44	85	59	4 964	3 492	3 251
...	0.03	0.03	21.61	20.46	19.72
119	103	140	13 017	12 811	15 660
7[a]	14	...	383[a]	578	562[b]
...	...	5[e]	575[a]	693	654[e]
...	...	...	413[a]	443[d]	516[b]
45/71	...	...	...	...	...
63/48[a]	71/56	75/59[e]	94/95[a]	97/97[d]	104/104[b]
3.2[a]	2.6[d]	...	7.0[a]	6.9[d]	6.7[b]
...	13 681[r]	...	580	550[r]	...
2 326	2 173	2 074	3 398	3 401	3 443

a: 1975; **b:** 1985; **c:** 1982; **d:** 1980; **e:** 1984; **f:** Inter-trade between the members the Customs and Economic Union of Central Africa is excluded; **h:** Imports f.o.b.; Yaoundé; **n:** 1983; **p:** Industrial products; (manufacturing industry only), including pr duction of exported products; **r:** 1979; **s:** Currency: CFA francs; **t:** Currency: Can dian dollars.

Cape Verde			Central African Republic		
1976	1981	1986	1976	1981	1986
0.30[a]	0.30[a]	0.33[a]	2.10	2.35[a]	2.74
46.8	46.0	41.1	40.6	41.6	42.5
7.4/8.8	7.8/8.7	6.3/7.3	5.8/6.9	5.6/6.9	5.5/6.7
54/5	...	...	58/52	...	...
...	...	...	...	...	...
...	...	...	...	...	...
5.1	5.1	5.3	34.2	38.2	42.4
0.9	1.9	2.4	2.2	2.3	2.4
55/58	57/61	60/63	40/44	41/45	43/47
87	75	63	145	142	132
...	...	...	...	797[h]	641[c]
...	...	...	...	2.7	1.4
...	...	...	...	347[h]	233[c]
...	20	19[k]	...	...	...
...	20	23[k]	...	...	...
...	9	9[k]	...	...	...
...	5	5[k]	...	...	...
...	...	...	93	103	108
...	...	...	99	101	94
...	...	...	...	...	...
...	...	...	4[b]	5	7
30	68[h]	...	55[t]	95[t]	109[ct]
2	4[h]	...	59[t]	79[t]	88[ct]
...	100.0[e]	130.0[e]	...	100.0	150.3[f]
...	...	...	...	115[g]	168[g]
31.549[r]	50.860[r]	76.565[r]	248.50[s]	287.40[s]	322.750[s]
...	...	...	19	69	65
...	...	...	0.00	0.01	0.01
...	...	...	...	7	4
9[b]	...	11[k]	6[b]	8	18[n]
6[b]	6	...	...	...	3[e]
...	...	8[k]	...	0.3[h]	1.7[k]
...	...	39/61[c]	67/85[b]	...	...
78/70[b]	67/61[h]	68/64[c]	55/27[b]	57/30[h]	44[p]
3.9	...	2.6[c]	4.2[b]	3.7[h]	5.4[c]
6 670[q]	6 353[h]	...	21 430	22 484[h]	...
2 177	2 545	2 614	2 240	2 115	2 045

: United Nations estimate; **b:** 1975; **c:** 1985; **e:** Base: 1983 = 100; Praya; excluding ent; **f:** Excluding rent; **g:** Bangui, including exported products; excluding products f mining and quarrying; **h:** 1980; **k:** 1984; **n:** 1983; **p:** Both series; **q:** 1977; **r:** Curreny: Escudo; **s:** Currency: CFA francs; **t:** Inter-trade between the members of the ustoms and Economic Union of Central Africa is excluded.

Chad

1976	1981	1986
4.12	4.58[a]	5.14[a]
41.7	41.9	42.3
5.3/6.4	5.3/6.3	5.3/6.3
61/18[b]	...	...
...	...	...
...	...	...
15.2	20.8	27.0
2.1	2.3	2.4
39/43	41/45	43/47
154	143	132
758[b]	...	...
2.6	...	...
188[b]	...	...
17	14	12[d]
38	45	58[d]
14	9	10[d]
13	8	8[d]
95	96	120
98	94	106
...	...	...
...	...	...
116	108	162[d]
63	83	138[d]
...	...	...
...	...	...
248.5[s]	287.4[s]	322.75[s]
23	7	16
0.00	0.01	0.01
...	...	25
...	...	...
...	...	1[c]
...	...	...
...	...	...
19[bp]	...	33/12[d]
2.2[b]	1.9[k]	...
47 889[b]	...	...
...	...	...

Chile

1976	1981	1986
10.37	11.33	12.33
36.0	32.3	30.2
6.7/8.4	7.0/8.9	7.2/9.4
49/15[b]	46/19	47/20[d]
...	22.0	14.6[d]
...	17.1	16.4[d]
78.3	81.1	83.6
1.5	1.6	1.5
65/71	67/73	68/74
46	23	20
4 170[b]	27 571[h]	15 996[c]
-2.3	7.9	-1.1
403[b]	2 478[h]	1 329[c]
13	19	14[c]
8	6	6[k]
36	30	34[k]
23	22	21[k]
86	106	116
92	104	106
74[e]	101[e]	106[e]
3 890	4 553	4 851
1 684	6 364	3 157
2 083	3 906	4 222
20.6[f]	119.7[f]	313.4[f]
...	109[g]	363[g]
17.42[t]	39.00[t]	204.73[t]
405	3 213	2 351
1.34	1.70	1.53
235	405	547
25[b]	66	73[c]
42[b]	52	63[c]
68[b]	110[h]	145[c]
...	8/9[n]	...
92/94[b]	91/93[n]	93/94[d]
3.9	4.4[h]	4.4[c]
2 117[f]	1 926[r]	...
2 601	2 642	2 589

a: United Nations estimate; **b:** 1975; **c:** 1985; **d:** 1984; **e:** Manufacturing; **f:** Santiag
g: Domestic supply; **h:** 1980; **k:** 1983; **n:** 1982; **p:** Both sexes; **r:** 1979; **s:** Currency
CFA francs; **t:** Currency: Chilean pesos.

China 1976	China 1981	China 1986	Colombia 1976	Colombia 1981	Colombia 1986
949.59[a]	1 011.22[a]	1 072.22[a]	24.33	26.43	29.19
39.5	35.2	29.7	42.6	39.4	37.2
6.1/7.8	6.6/8.2	7.5/9.0	4.7/5.6	5.0/6.1	5.3/6.5
57/36	...	...	...	50/16	...
...	...	...	...	28.5	...
...	...	...	...	14.5	...
20.2	20.4	20.6	60.8	64.2	67.4
2.2	1.4	1.2	2.1	2.1	2.0
63/64	66/66	67/69	60/65	61/66	63/67
61	41	39	55	50	46
134 570[b]	247 185[h]	239 028[k]	13 098[b]	33 395[h]	34 187[c]
5.2[s]	7.0[s]	9.8[s]	6.0	5.6	2.1
148[b]	253[h]	234[k]	565[b]	1 295[h]	1 191[c]
31[r]	28[r]	30[kr]	16	18	17[p]
41	42	45[k]	24	19	18[p]
43	43	42[k]	27	25	28[p]
...	...	...	24	21	22[p]
84	103	134	87	103	107
89	101	125	97	100	97
...	...	...	83	100	...
334 480	418 487	581 558	12 921	15 817	28 743
6 580	22 014	42 620	1 708	5 199	3 464
6 860	22 007	31 050	1 745	2 956	5 102
88.3	102.5	131.2	41.2[d]	129.4[d]	329.6[d]
...	...	...	...	124[e]	309[e]
1.880[t]	1.745[t]	3.722[t]	36.32[fx]	59.07[fx]	219.00[fx]
2 345[g]	5 058	11 453	1 101	4 801	2 696
12.80[g]	12.70	12.70	1.41	3.37	2.01
...	7 767	9 000	477	1 060	732
0.1[b]	1.0	...	16[b]	31	40[k]
...	4	6[c]	52[b]	65	70[c]
1[b]	4[h]	9.4[c]	69[b]	87[h]	96[c]
...	21/49[n]	...	...	14/16	...
96/80[b]	88/71[h]	85/69[c]	76/79[b]	81/83[h]	81/83[c]
1.7[b]	2.2[h]	2.4[c]	1.9[b]	1.7[h]	2.7[c]
1 492[b]	1 318	1 243[p]	1 810[b]	1 710	...
2 219	2 288	2 564	2 332	2 505	2 578

United Nations estimate; **b:** 1975; **c:** 1985; **d:** Bogota; low income group; **e:** Domestic pply; including exported products; **f:** Selling rate; **g:** 1977; **h:** 1980; **k:** 1983; **n:** 1982; 1984; **r:** Net fixed capital formation; **s:** Net material product; **t:** Currency: Yuan renminbi; **x:** Currency: Colombian pesos.

Comoros

1976	1981	1986
0.33[a]	0.39[a]	0.48
45.4	45.7	46.1
4.3/4.9	4.2/4.9	4.2/4.9
...	44/16[d]	...
...	53.3[d]	...
...	4.2[d]	...
21.3	23.2	25.2
3.4	3.0	3.1
46/50	48/52	50/54
97	88	80
...	...	...
...	...	...
...	...	...
...	23	27[g]
...	48	45[g]
...	6	6[g]
...	5	5[g]
...	...	...
...	...	...
...	...	...
...	...	...
23[b]	34	...
10[b]	16	...
...	...	...
...	...	...
248.49[p]	287.40[p]	322.75[p]
...	...	...
...	...	...
...	3	5
...	...	...
...	...	...
...	...	...
...	44/60[d]	...
...	72/48[d]	...
...	4.9[h]	...
...	18 200[k]	...
2 092	2 074	2 090

Congo

1976	1981	1986
1.39	1.57	1.79
42.8	43.2	43.6
4.9/5.8	4.9/5.8	5.0/5.8
...	...	...
...	...	...
...	...	...
35.8	37.3	39.5
2.5	2.6	2.7
43/46	45/48	47/50
85	81	73
...	1 706	2 194[g]
...	...	...
...	1 116	1 189[g]
30	44	29[g]
13	8	7[g]
34	46	49[g]
8	6	4[g]
92	104	108
101	101	92
...	...	...
2 120	4 292	5 871
168[e]	804[e]	751[e]
181[e]	1 073[e]	1 077[e]
68.4[f]	117.1[f]	170.1[c]
...	...	...
248.49[p]	287.40[p]	322.75[p]
12	123	7
0.06	0.01	0.01
...	36	39
20[b]	24	35[g]
...	11	...
2[b]	2.3[d]	2.9[c]
...	...	29/45[c]
...	...	...
6.7	6.4[d]	5.0[g]
6 460	5 307[k]	...
2 271	2 443	2 532

a: United Nations estimate; **b:** 1975; **c:** 1985; **d:** 1980; **e:** Inter-trade between t members of the Customs and Economic Union of Central Africa is excluded; **f:** Bra zaville; Europeans; excluding rent; **g:** 1984; **h:** 1982; **k:** 1978; **p:** Currency: CFA franc

Costa Rica			Côte d'Ivoire		
1976	1981	1986	1976	1981	1986
2.01	2.27	2.67[b]	7.04[a]	8.48[a]	10.16[a]
42.0	38.5	36.7	43.8	44.8	45.6
4.9/5.5	5.1/5.9	5.4/6.3	4.7/5.2	4.6/5.0	4.6/5.8
...	52/17	53/19[d]	...	...	...
...	26.9	31.4[d]	...	...	...
...	16.2	14.7[d]	...	...	...
42.2	46.0	50.0	32.2	37.1	42.0
3.0	2.6	2.4	3.8	3.7	3.5
69/74	71/76	71/76	46/50	49/52	51/54
30	20	18	121	110	100
1 961[c]	4 832[h]	3 814[d]	3 894[c]	10 176[h]	...
6.4	5.6	0.6	6.4	6.4	...
998[c]	2 120[h]	1 467	575[c]	1 245[h]	...
23	24	19[d]	22	24	12[n]
20	23	20[d]	26	29	28[n]
22	21	26[d]	13	13	16[n]
...	...	...	12	10	12[n]
91	101	111	78	109	118
108	96	93	85	101	98
...	...	...	...	...	...
123	191	247	29	537	1 128
770	1 209	1 130	1 297	2 384	1 742[d]
593	1 008	1 026	1 631	2 535	2 939[d]
70.2[f]	137.1[f]	497.8[f]	51.8[e]	108.5[e]	139.8[e]
...	165[g]	516[dg]	...	...	...
8.57[q]	36.090[q]	58.875[q]	248.49[r]	287.40[r]	322.75[r]
95	131	523	76	18	20
0.43	0.03	0.07	0.00	0.04	0.04
299	333	261	122	195	187
30[c]	63	68[k]	11[c]	26	...
63[c]	109	123[d]	9[c]	13[h]	...
65[c]	71[h]	77[d]	16[c]	37[h]	51[d]
...	...	7/7[n]	...	...	...
78/79[c]	77/80[h]	75/75[d]	49/30[c]	62/39[h]	61/39[n]
6.4[c]	7.1[h]	4.7[d]	5.6	6.0	...
1 524	1 441[p]	...	15 234	...	...
2 563	2 621	2 772	2 322	2 569	2 448

United Nations estimate; **b:** Marked break in series; **c:** 1975; **d:** 1985; **e:** Abidjan; San José; metropolitan area; **g:** San José; domestic supply: excluding products of ining and quarrying; **h:** 1980; **k:** 1983; **n:** 1984; **p:** 1979; **q:** Currency: Colones; **r:** urrency: CFA francs.

Cuba 1976	Cuba 1981	Cuba 1986	Cyprus 1976	Cyprus 1981	Cyprus 1986
9.43	9.72	10.25	0.61	0.63	0.67
36.9	31.3	26.4	25.9	24.4	25.4
9.9/9.2	10.6/10.3	11.0/11.1	13.1/14.7	12.8/15.2	12.3/14.6
47/12	49/23	...	...	...	...
...	22.3	...	...	21.4	17.4[c]
...	18.9	...	...	20.0	18.7[c]
64.2	68.1	71.8	43.4	46.3	49.5
0.8	0.6	1.0	0.7	1.2	1.0
71/74	72/75	72/75	72/76	72/76	73/77
23	17	15	23	17	15
9 954[ah]	13 781[gh]	15 152[bh]	698[a]	2 137[g]	2 337[b]
7.9	3.7	8.7	-3.3	11.1	5.3
1 067[a]	1 416[g]	1 520[b]	1 146[a]	3 397[g]	3 493[b]
19[k]	20[k]	22[bk]	21	32	28[b]
7	13	10[b]	16	9	7[b]
45	31	36[b]	21	20	19[b]
...	...	...	17	18	16[b]
82	105	113	94	99	96
82	103	108	97	98	89
...	...	...	70	104	112
258	266	948	...	...	...
4 218	6 545	9 173	432	1 165	1 279
3 571	5 406	6 298	258	556	506
...	...	...	74.9[d]	110.8	139.6
...	...	...	...	110[e]	125[e]
0.830[p]	0.830[p]	0.793[p]	0.412[q]	0.433[q]	0.512[q]
...	...	...	275	426	753
...	...	...	0.43	0.46	0.46
...	121	162[f]	180	421	901
15[a]	18	37[b]	107[a]	227	258[b]
32[a]	41	51[b]	116[a]	202	405[b]
64[a]	131[g]	197[b]	...	137[g]	132[b]
...	4/4	...	15/3	...	...
87/89	94/94[g]	93/94[b]	...	...	...
...	6.5[g]	5.8[b]	3.9[a]	3.3[g]	3.7[b]
996[a]	638[g]	...	1 220	1 060	...
2 652	2 834	3 094	...	...	...

a: 1975; **b:** 1985; **c:** 1984; **d:** 1977; **e:** Domestic supply; **f:** 1983; **g:** 1980; **h:** Net mater product; **k:** Net fixed capital formation; **p:** Currency: Cuban pesos; **q:** Currency: Cypri pounds.

Czechoslovakia			Democratic Kampuchea		
1976	1981	1986	1976	1981	1986
14.92	15.32	15.53	6.97[a]	6.46[a]	7.49[a]
23.4	24.3	24.5	41.6	32.9	32.5
5.0/19.7	13.2/18.1	13.6/18.8	4.3/5.2	4.0/4.7	4.1/5.1
56/44	56/47[d]	...	47/32	...	...
...	13.1[d]	...	...	...	...
...	39.3[d]	...	...	...	...
59.1	62.3	67.7	10.3	10.8	11.6
0.7	0.3	0.3	-2.1	2.6	2.5
67/74	67/75	68/76	30/32	42/45	47/50
19	16	14	263	160	130
412[kn]	470[kn]	549[ckn]	...	...	...
5.4[n]	3.7[n]	1.7[n]	...	...	...
27 613[p]	30 678[p]	85 351[cp]	...	...	...
21[h]	18[h]	17[ch]	...	...	...
8	7	8[c]	...	...	...
67	61	61[c]	...	...	...
...	...	...	...	...	...
91	102	119	143	104	165
93	102	117	130	103	142
84	102	118	...	...	...
45 231	45 128	47 121	...	3	6
9 706[e]	14 658[e]	21 055[e]	...	...	...
9 035	14 876	20 456	...	...	...
90.9	100.8	111.0	...	...	...
...	...	...	...	...	...
11.46[q]	11.94[q]	9.71[q]	...	...	...
...	...	...	...	...	...
...	...	...	...	...	...
...	4 787	5 330	...	...	...
102[b]	182	201[c]	...	...	...
176[b]	210	231[c]	...	...	...
249[bf]	280[df]	280[cf]	4.2[b]	5.5[d]	7.1[c]
...	...	...	...	...	...
74/80[b]	74/82[d]	77/84[c]	...	...	...
4.3[b]	4.5[d]	4.9[c]	...	...	...
420	345[g]	...	...	...	...
3 434	3 434	3 479	...	...	...

: United Nations estimate; **b:** 1975; **c:** 1985; **d:** 1980; **e:** Imports f.o.b.; **f:** Number f licences issued or sets declared; **g:** 1982; **h:** Net fixed capital formation; **k:** Billion Koruny; **n:** Net material product; **p:** In national currency; **q:** Currency: Koruny.

Democratic People's Republic of Korea			Democratic Yemen		
1976	1981	1986	1976	1981	1986
16.25[a]	18.48[a]	20.88[a]	1.74	2.03	2.36
41.6	40.3	38.7	47.5	45.9	45.1
5.1/6.1	4.8/6.3	4.7/6.7	4.3/4.6	4.3/4.8	4.4/4.9
48/40[b]	...	...	...	...	...
...	...	...	...	...	...
...	...	...	...	...	...
55.1	59.7	63.8	34.3	36.9	39.9
2.6	2.5	2.4	2.4	2.8	3.0
62/69	65/71	66/73	45/47	47/50	49/52
35	30	24	150	135	120
...	...	...	262[b]	668[g]	...
...	...	...	-7.8	14.7	...
...	...	...	158[b]	359[g]	...
...	...	...	40	37[g]	...
...	...	...	16	10	...
...	...	...	13	13	...
...	...	...	11	11	...
86	103	123	103	100	99
95	100	106	112	98	83
...	...	...	...	...	...
27 848	30 831	35 394	...	...	...
...	...	...	412	703	483
...	...	...	177	22	29
...	...	...	73.8[e]	103.8[e]	127.8[de]
...	...	...	...	...	...
2.06[bfh]	2.02[fh]	2.24[fh]	0.345[k]	0.345[k]	0.345[k]
...	...	...	82	255	138
...	...	...	0.02	0.04	0.04
...	...	...	...	...	...
...	...	...	7[b]	14	24[c]
...	...	...	...	...	...
...	...	10[c]	19[b]	19[g]	18[c]
...	...	...	...	...	...
...	...	...	73/29[b]	73/27	...
...	...	...	...	5.7	...
...	...	...	9 090	7 632[g]	...
2 771	3 060	3 131	1 910	2 211	2 293

a: United Nations estimate; **b:** 1975; **c:** 1985; **d:** 1984; **e:** Aden; **f:** Operational e change rate for United Nations programmes as of 1 October; **g:** 1980; **h:** Currenc Won; **k:** Currency: Yemeni dinars.

Denmark 1976	Denmark 1981	Denmark 1986	Djibouti 1976	Djibouti 1981	Djibouti 1986
5.07	5.12	5.12	0.23[a]	0.37	0.46
22.6	20.8	18.7	...	...	...
6.8/20.6	17.2/21.6	17.7/22.5	...	...	...
58/40	59/46	59/48[c]	...	...	...
9.3	6.7	6.4[c]	...	...	...
23.8	22.3	20.9[c]	...	...	...
82.1	84.3	85.9	68.5	73.7	76.1[n]
0.5	0.3	-0.0	6.0	7.7	3.4[n]
71/77	71/78	72/78	...	...	...
9	8	7	...	...	...
37 636[b]	66 321[g]	58 062[c]	...	339[g]	...
2.2	2.4	2.6	...	-1.9	...
7 438[b]	12 946[g]	11 336[c]	...	1 094[g]	...
23	16	19[c]	8[p]	13	...
5	5	5[c]	3	3	...
19	18	19[c]	9	9	...
18	16	17[c]	6	7	...
86	101	126	...	...	...
87	101	126	...	...	...
93	100	126	...	...	...
197	761	5 581	...	...	...
12 427	17 803	22 871	...	...	...
9 115	16 286	21 276	...	...	...
66.4	111.7	151.7	...	107.7[d]	104.1[d]
...	115[e]	138[e]	...	...	...
5.787[q]	7.325[q]	7.342[q]	177.72[bfr]	177.00[fr]	177.00[fr]
842	2 548	4 965	...	...	...
1.81	1.63	1.63	...	...	...
...	3 570	...	...	...	...
257[b]	317	350[c]	53[b]	55	61[c]
454[b]	674	783[c]	14[b]	13	19[c]
308[bh]	362[gh]	386[ch]	15[b]	16[g]	27[c]
...	...	...	...	...	...
92[bk]	101/100[g]	101/100[n]	...	...	...
6.8[b]	6.1[g]	...	...	...	...
533	482[p]	399[c]	...	2 196	...
3 342	3 585	3 529	...	...	...

: United Nations estimate; **b:** 1975; **c:** 1985; **d:** Europeans; excluding rent; **e:** Domestic upply; including production of exported and agricultural products and products of ne manufacturing industry; **f:** Operational exchange rate for United Nations programnes as of 1 October; **g:** 1980; **h:** Number of licences issued or sets declared; **k:** Both exes; **n:** 1984; **p:** 1978; **q:** Currency: Danish kroner; **r:** Currency: Djibouti francs.

Dominica			Dominican Republic		
1976	1981	1986	1976	1981	1986
0.07	0.08	0.08[a]	4.89	5.58	6.42
...	...	...	47.0	43.9	40.7
...	...	...	4.4/4.4	4.5/4.4	4.7/4.7
...	45/23	...	46/16[d]	48/20	...
...	31.0	...	...	22.0	...
...	6.6	...	...	12.6	...
...	...	...	45.3	50.5	55.7
0.6	0.3	1.3	2.3	2.3	2.2
...	...	...	58/62	61/65	63/64
...	...	...	84	75	65
29[b]	59[p]	85[h]	3 599[b]	6 631[p]	4 651[c]
-3.9	1.7	5.1	9.3	4.5	1.7
403[b]	808[p]	1 755[h]	728[b]	1 193[p]	745[c]
23	32	39[h]	22	23	20[h]
34	27	25[h]	19	19	18[h]
7	9	10[h]	25	20	17[h]
4	6	7[h]	21	16	16[h]
...	...	...	97	99	106
...	...	...	105	98	97
...	...	...	78	97	116[ce]
1	1	1[h]	...	69	77
21[b]	47[f]	...	764[g]	1 450[g]	1 433[g]
11[b]	24[f]	...	716	1 188	718
56.1	117.3	130.4[h]	...	107.5[k]	232.5[k]
...	...	...	...	...	...
2.70[s]	2.70[s]	2.70[s]	1.00[nt]	1.00[nt]	3.08[nt]
...	...	...	124	225	376
...	...	...	0.09	0.14	0.02
...	16	24	260[r]	340[r]	792[r]
28[b]	...	...	15[b]	28	24[c]
37[b]	...	77[h]	24[b]	30	...
...	...	...	36[b]	72[p]	80[c]
...	...	...	...	32/31	...
...	...	...	74[bq]	81[pq]	83/92[c]
...	...	...	1.6	...	1.9[h]
...	8 100[d]	...	4 023	...	...
2 167	2 379	2 615	2 234	2 316	2 468

a: United Nations estimate; **b:** 1975; **c:** 1985; **d:** 1978; **e:** Manufacturing; **f:** 1982; **g** Imports f.o.b.; **h:** 1984; **k:** Including indirect taxes; **n:** Fixed rate; **p:** 1980; **q:** Bo sexes; **r:** Arrivals by air only; **s:** Currency: East Caribbean dollars; **t:** Currenc Dominican Republic pesos.

Ecuador			Egypt		
1976	1981	1986	1976	1981	1986
7.24	8.36	9.65	37.87	43.31	49.61
44.7	43.3	41.8	40.0	40.0	39.6
5.1/5.9	5.1/5.8	5.2/5.8	5.7/6.8	5.7/6.8	5.6/6.8
...	47/12[c]	...	51/5[d]	48/6[c]	50/12[e]
...	33.0[c]	...	47.8	36.8[c]	38.3[e]
...	12.9[c]	...	14.6	15.4[c]	14.8[e]
42.4	47.3	52.3	43.7	44.7	46.4
2.9	2.9	2.8	2.7	2.4	2.3
60/63	62/66	63/68	54/57	57/59	59/62
82	70	63	120	100	85
4 310[a]	11 733[p]	15 982[b]	13 418[a]	24 499[p]	...
12.7	6.4	1.7	5.1	9.2	...
613[a]	1 444[p]	1 704[b]	370[a]	590[p]	...
22	22	17[b]	22	30	...
17	29	17[b]	26	18	...
12	31	17[b]	20	27	...
14	36	19[b]	15	13	...
94	104	122	94	101	115
105	101	101	107	99	104
70	102	117[f]	53	117	...
9 592	10 886	15 116	17 909	32 435	46 016
959	2 246	1 867	3 808	8 839	11 502
1 258	2 542	2 171	1 522	3 233	2 934
...	100[g]	056.7[g]	602	110.5	239.3
...	...	...	...	107[h]	192[h]
25.00[r]	25.00[r]	146.50[r]	0.39[ks]	0.70[ks]	0.70[ks]
477	632	644	240	716	829
0.39	0.41	0.41	2.43	2.43	2.43
172	245	306	984[n]	1 376[n]	1 311[n]
7[a]	29	28[q]	6[a]	14	24[q]
26[a]	33	36[b]	13[a]	12	24[b]
36[a]	62[p]	64[b]	17[a]	34[p]	82[b]
...	16/24[c]	...	43/71	...	...
75/72[a]	85/84[p]	...	73/46[a]	78/53[p]	84/65[b]
...	5.3[p]	3.3[b]	4.3[a]	3.6	4.9[b]
2 010[a]	...	...	1 147[a]	815	...
2 037	2 063	2 031	2 692	3 027	3 262

a: 1975; **b:** 1985; **c:** 1982; **d:** 1977; **e:** 1983; **f:** Manufacturing; **g:** Index base: 1981 = 100; **h:** Domestic supply; excluding products of mining and quarrying; **k:** Fixed rate; **n:** Visitor arrivals; **p:** 1980; **q:** 1984; **r:** Currency: Sucres; **s:** Currency: Egyptian pounds.

El Salvador / Equatorial Guinea

El Salvador 1976	El Salvador 1981	El Salvador 1986	Equatorial Guinea 1976	Equatorial Guinea 1981	Equatorial Guinea 1986
4.12	4.59	4.91	0.32[a]	0.36[a]	0.40[a]
45.7	45.2	44.6	39.9	40.7	41.4
4.7/5.7	4.6/5.6	4.6/5.7	6.3/7.3	6.2/7.2	6.0/7.0
46/18	48/24	...	58/3	...	...
47.2	40.0	...	...	...	...
10.5	16.4	...	...	...	...
39.4	39.3	39.1	46.6	53.7	59.7
2.9	2.9	3.1	2.0	2.1	2.3
60/65	63/67	65/69	40/44	42/46	44/48
82	70	59	149	137	127
1 791[b]	3 567[f]	5 732[c]	...	...	90[c]
5.5	1.8	-1.5	...	...	...
432[b]	744[f]	1 032[c]	...	...	230[c]
20	14	12[c]	...	12	15[c]
28	24	18[c]	...	44	59[c]
18	18	19[c]	...	7	3[c]
16	16	16[c]	...	5	1[c]
85	90	85	...	...	...
102	91	84	...	...	...
...	...	...	...	...	...
40[b]	113	144	0	0	0
718	985	961[c]	...	...	...
721	797	679[c]	...	...	...
...	114.8[d]	261.6[d]	...	...	...
...	...	...	...	...	...
2.5[ep]	2.5[ep]	5.0[ep]	682.88	194.90	322.75[q]
185	72	170	...	...	...
0.49	0.52	0.47	...	...	...
278	82	134	...	...	...
...	24	...	...	...	...
15[b]	22	24[g]	...	...	...
33[b]	63[f]	63[c]	1.6[b]	2.8[f]	5.6[c]
38/34[b]	33[fh]	...	...	63[fh]	...
64/62[b]	63/63[f]	59/61[g]	50[bh]	56[fh]	...
3.1[b]	3.6[f]	2.6[g]	...	...	...
3 681[k]	3 184[f]	2 718[g]	63 800[b]	...	...
...	...	...	...	...	...

a: United Nations estimate; **b:** 1975; **c:** 1985; **d:** Urban areas; **e:** Buying rate; **f:** 1980; **g:** 1984; **h:** Both sexes; **k:** 1977; **p:** Currency: Colones; **q:** Currency: CFA francs. Prior to January 1985, national currency relates to Bipkwele.

Ethiopia

1976	1981	1986
28.19	39.44	44.93
44.4	44.4	44.8
4.0/4.7	4.0/4.7	4.0/4.7
57/37d	55/35e	54/34r
...	...	...
...	...	...
9.5	10.5	11.6
2.3	2.5	2.8
39/42	39/42	40/43
155	155	149
2 669b	4 072e	...
1.2	2.8	...
78b	106e	...
10	10	13r
46	45	41r
11	10	11r
10	10	10r
87	99	103
94	96	88
...	...	...
30b	43	56
356	739	1 102
292	389	455
61.9h	106.1h	100.0h
...	...	...
2.07s	2.07s	2.07s
295	267	251
0.27	0.26	0.21
37n	46n	59n
1.9b	1.6	1.4c
2b	28e	30c
0.6b	0.8e	1.6c
...	38q	...
18/9b	30/16e	30/20c
2.7	2.9	3.1c
73 184d	72 582e	...
...	...	...

Fiji

1976	1981	1986
0.59	0.65	0.70a
39.9	37.5	37.2
4.6/4.4	5.1/5.0	5.4/5.6
49/10	52/13g	...
43.8	...	...
9.2	...	...
36.7	38.7	41.2
1.8	1.9	1.6
66/69	67/71	68/73
37	31	26
683b	1 204e	1 162c
6.5	4.4	1.0
1 186b	1 914e	1 682c
19	27	17c
...	18	16c
...	11	13c
...	9	9c
72	103	120
77	101	108
80	110	123
...	...	29
263	632	422
135	311	264
71.4	111.2	101.8k
...	...	...
0.942t	0.877t	1.145t
116	135	171
...	0.01	0.01
169	206	258
29b	73	79c
50b	71	75c
...	...	...
16/26	...	...
90/90b	94/95e	94/95c
4.3b	5.5	...
2 391	2 232e	...
2 643	2 774	2 932

United Nations estimate; **b:** 1975; **c:** 1985; **d:** 1977; **e:** 1980; **g:** 1982; **h:** Addis Ababa; excluding rent; **k:** Base: 1985 = 100; **n:** Air arrivals; **q:** 1983; **r:** 1984; **s:** Currency: Birr; **t:** Currency: Fiji dollars.

Finland 1976	Finland 1981	Finland 1986	France 1976	France 1981	France 1986
4.73	4.80	4.92	52.89	54.18	55.39
22.0	20.3	19.3	23.9	22.3	21.3
12.4/18.4	12.7/19.8	13.4/20.8	15.2/21.2	14.2/20.1	14.9/20.4
52/39	51/41[c]	58/49[b]	54/30[a]	51/42	53/35[d]
14.3[a]	12.6[c]	11.2[b]	9.6[a]	8.2	7.0[d]
26.4[a]	26.3[c]	24.3[b]	28.4[a]	25.1	22.7[d]
55.1	59.6	64.0	73.0	73.2	73.4
0.3	0.5	0.3	0.4	0.3	0.3
68/77	70/78	71/79	70/78	71/79	71/79
9	6	6	11	9	8
28 348[a]	51 624[c]	54 113[b]	338 852[a]	655 305[c]	510 333[b]
4.6	3.2	2.7	4.2	3.3	1.2
6 017[a]	10 800[c]	10 959[b]	6 429[a]	12 200[c]	9 343[b]
28	25	23[b]	23	21	19[b]
9	8	7[b]	5	4	4[d]
27	28	26[b]	30	28	29[d]
24	24	23[b]	27	25	25[d]
109	93	112	90	99	107
110	93	108	92	98	105
79	103	117	92	98	101
966	2 586	4 124	31 232[e]	38 627[e]	45 729[e]
7 392	14 202	15 325	64 383	120 872	128 835
6 341	14 015	16 340	55 699	101 371	119 430
69.0	112.0	155.9	66.7	113.4	162.2
...	113[f]	133[f]	...	111[g]	161[g]
3.767[p]	4.357[p]	4.794[p]	4.970[q]	5.748[q]	6.455[q]
462	1 484	1 787	5 620	22 262	31 454
0.82	1.27	1.91	101.02	81.85	81.85
264	380	452[b]	25 036	31 340	36 080
212[a]	303	354[b]	290[a]	417	438[b]
389[a]	522	617[b]	262[a]	498	608[b]
352[a]	414[c]	470[b]	...	354[c]	394[b]
...	...	...	...	...	...
93/97[a]	96/103	99/107[b]	93[ah]	93/98	103[b]
5.6[a]	5.4	5.1[d]	4.6[a]	5.2	...
703[a]	504	452[d]	650	480[k]	...
3 141	3 075	3 008	3 439	3 321	3 327

a: 1975; **b:** 1985; **c:** 1980; **d:** 1984; **e:** Including Monaco; **f:** Domestic supply; **g:** dustrial products; manufacturing industry only exclusive of finished goods; **h:** Bo sexes; **k:** 1982; **p:** Currency: Markkaa; **o:** Currency: Francs.

abon 1976	1981	1986	Gambia 1976	1981	1986
1.01[a]	1.08[a]	1.17[a]	0.54	0.62	0.66[a]
33.2	33.4	34.6	41.7	42.1	42.5
8.6/10.0	8.7/10.1	8.7/10.1	4.8/5.7	4.8/5.6	4.6/5.4
61/37	...	...	56/44	51/44[c]	...
...	...	...	...	73.7[c]	...
...	...	...	...	2.9[c]	...
30.6	35.8	40.9	16.6	18.1	20.1
1.2	1.6	2.0	2.1	1.9	2.1
45/49	47/51	49/53	32/35	33/36	35/39
122	112	103	185	174	164
2 158[b]	...	...	143[b]	222[h]	...
20.6	...	...	6.2	-1.9	...
2 152[b]	...	...	273[b]	380[h]	...
61	...	...	...	...	...
5	...	...	40	35	...
37	...	...	6	6	...
5	...	...	5	6	...
91	100	107	123	120	135
95	99	97	137	118	121
...	...	...	...	...	...
11 494	11 348	8 261	...	...	...
501[e]	841[e]	976[de]	75	122	100
1 136[e]	2 200[e]	1 920[de]	37	27	35
65.4[f]	108.7[f]	159.3[bf]	72.2[g]	106.1[g]	188.0[dg]
...	...	...	...	...	...
248.50[p]	287.40[p]	322.75[p]	1.98[q]	2.096[q]	7.43[q]
116	199	126	21	4	14
...	0.01	0.01	...	...	...
...	15	28[d]	32	22	75
...	...	...	3[b]	10	10[k]
...	...	11[c]	5[b]	...	...
...	9[h]	19[d]	...	...	...
...	...	...	...	...	...
63/55[b]	71/64[h]	...	31/14[b]	46/23[h]	63/37[d]
1.8[b]	2.0[h]	3.3[d]	2.5[b]	4.0	4.1[k]
...	4 971[h]	...	13 020	...	...
...	...	...	2 133	2 176	2 229

United Nations estimate; **b:** 1975; **c:** 1983; **d:** 1985; **e:** Inter-trade between the embers of the Customs and Economic Union of Central Africa is excluded; **f:** breville; **g:** Banjul, Kombo, St. Mary; **h:** 1980; **k:** 1984; **p:** Currency: CFA francs; Currency: Dalasis.

German Democratic Republic			Germany, Federal Republic of		
1976	1981	1986	1976	1981	1986
16.79	16.74	16.62	61.51	61.67	61.05
21.6	17.9	17.9	21.8	18.2	15.4
17.4/25.6	14.1/23.4	13.0/22.9	16.2/23.6	14.8/23.4	15.0/24.6
51/44a	53/46	55/48	57/31	59/33	60/35d
10.9	10.7	10.8	6.4a	5.7c	5.1d
38.3	38.1	37.7	37.1a	36.3c	33.0d
75.4	76.4	76.6	83.1	84.4	85.5
-0.3r	-0.1	-0.1	-0.1	-0.2	-0.2
69/74	69/75	70/75	69/76	70/77	71/78
14	12	9	15	11	9
163 618st	202 971st	252 210st	417 439a	813 498c	624 969b
5.5t	4.2t	4.5t	2.4	3.3	1.3
9 747t	12 128t	15 171t	6 751a	13 213c	10 266b
35.3a	33.9c	26.1	20	22	20b
14.9	12.4	11.3	3	2	2b
58.4a	61.6c	63.9	39	35	32b
...	...	...	35	32	32b
90	103	110	91	100	115
89	103	109	91	100	116
83	105	127	91	98	107
55 005	59 528	70 123	115 256	114 158	108 552
13 196f	20 181f	27 414f	88 426e	163 934e	189 484e
11 361	19 858	27 729	102 166e	176 043e	242 411e
99.5	100.2	100.1	85.6	106.3	120.7
...	...	...	...	114gh	113h
2.40v	2.20v	2.00v	2.362x	2.255x	1.94x
...	...	...	30 019	43 719	51 734
...	...	...	117.61	95.18	95.18
1 085	1 457	1 500d	9 356	11 120	12 217
143a	206	238b	290a	414	450b
153a	194	218b	317a	488	621b
310a	347n	363b	311a	337cn	373b
...	...	...	...	...	...
93p	86/84c	87/84b	79/81p	77/80c	79/80b
5.4a	5.3c	5.4b	4.1a	4.1c	4.2d
530b	483	425	500	431q	...
3 485	3 645	3 768	3 284	3 433	3 475

a: 1975; **b:** 1985; **c:** 1980; **d:** 1984; **e:** Excluding trade conducted in accordance w the supplementary protocol to the treaty on the basis of relations between the Fede Republic of Germany and the German Democratic Republic; **f:** Imports f.o.b.; **g:** 198
h: Domestic production; **n:** Number of licences issu or sets declared; **p:** Both sexes; **q:** 1982; **r:** 1971-76; **s:** Million national currency; Net material product; **v:** Currency: Mark; **x:** Currency: Deutsche mark.

Ghana			Greece		
1976	1981	1986	1976	1981	1986
10.31[a]	11.94[a]	14.04[a]	9.17	9.73	9.97
45.9	46.5	46.7	23.9	22.8	21.5
4.2/4.9	4.2/4.9	4.2/4.9	16.1/18.6	16.0/18.9	16.3/19.3
44/31	...	...	58/27[b]	54/19	...
...	...	...	32.7[d]	27.4	...
...	...	...	20.4[d]	20.1	...
29.8	30.7	31.5	55.3	57.7	60.1
3.3	3.2	3.4	1.3	0.5	0.4
49/53	50/54	52/56	71/75	72/76	73/77
103	98	90	25	16	15
2 882[b]	4 788[h]	6 900[c]	20 818[b]	40 147[h]	33 407[c]
1.2	1.6	-0.8	4.9	4.5	1.2
294[b]	414[h]	508[c]	2 301[b]	4 163[h]	3 382[c]
10	5	9[c]	21	22	19[c]
51	53	41[c]	17	16	15[c]
15	7	13[c]	21	21	21[c]
13	6	11[c]	18	18	16[c]
102	98	137	91	105	102
117	95	114	95	104	96
...	92	...	85	101	108
353	515	341	3 108	3 901	6 841
862	1 106	731[c]	6 087	8 798	11 314
827	1 063	617[c]	2 558	4 247	5 650
11.5	216.5	000.1[c]	53.3	124.5	314.8
...	150[e]	1 323[ce]	...	126[f]	[illegible][f]
1.15[gp]	2.75[gp]	90.01[gp]	37.03[q]	57.63[q]	138.76[q]
98	146	513	776	1 022	1 519
0.16	0.31	0.28	3.65	3.85	3.31
56	42	92	3 672	5 094	7 025
6[b]	9	...	49[b]	144	190[c]
6[b]	6	...	221[b]	302	355[k]
3.4[b]	5[h]	10[c]	116[b]	156[h]	174[c]
...	...	...	...	4/15	...
64/45[b]	64/47[h]	61/44[k]	96/87[b]	96/92	...
4.6[b]	...	1.4[k]	1.8[b]	2.2	...
10 640[b]	7 107	...	490[b]	394	...
2 201	1 785	1 679	3 405	3 571	3 660

United Nations estimate; **b:** 1975; **c:** 1985; **d:** 1977; **e:** Domestic supply; including ported products, excluding value added taxes; **f:** Domestic supply, finished products ly; excluding products of electricity, gas and water; **g:** Buying rate; **h:** 1980; **k:** 1984; Currency: Cedi; **q:** Currency: Drachmae.

Grenada 1976	Grenada 1981	Grenada 1986	Guatemala 1976	Guatemala 1981	Guatemala 1986
0.11	0.11[a]	0.11[a]	6.43	7.11	8.19
...	...	...	45.7	45.9	45.9
...	...	...	4.3/4.6	4.3/4.6	4.6/4.9
...	...	...	51/8[b]	48/8	...
...	...	...	58.2	54.0	...
...	...	...	13.9	11.1	...
...	...	...	37.1	38.5	40.0
3.1	0.4	1.2	2.8	2.8	2.9
...	...	...	55/58	57/61	60/64
...	...	...	82	70	59
34[b]	63[e]	96[c]	3 646[b]	7 879[e]	11 130[c]
-4.3	5.3	1.4	5.9	5.7	1.4
340[b]	589[e]	857[c]	605[b]	1 139[e]	1 398[c]
...	46	50[k]	21	17	11[c]
...	22	19[k]	...	...	...
...	10	10[k]	...	...	...
...	7	6[k]	...	...	...
...	...	...	91	102	101
...	...	...	104	101	95
...	...	...	...	...	...
...	...	...	42	237	347
25	54	83	839	1 673	576
13	19	28	760	1 226	624
...	100	148.0	666[d]	111.4[d]	162.5[c]
...	...	...	...	...	...
2.70[p]	2.70[p]	2.70[p]	1.00[qr]	1.00[qr]	2.50[c]
11	16	21	491	150	362
...	...	...	0.49	0.52	0.52
24	25	57	408	329	287
38[b]	...	...	13[b]	27	...
43[b]	54[e]	64[c]	...	14	16[c]
...	...	...	18[b]	25[e]	26[c]
...	...	...	...	...	37/53[c]
...	...	...	43/37[b]	50/43[e]	...
7.6[b]	...	5.6[c]	1.6	1.8	1.7[f]
4 360[b]	2 789	...	2 560	...	...
2 140	2 297	2 371	2 156	2 220	2 298

a: United Nations estimate; **b:** 1975; **c:** 1985; **d:** Urban areas; **e:** 1980; **f:** 1984;
1983; **p:** Currency: East Caribbean dollars; **q:** Currency: Quetzales; **r:** Buying ra

uinea 1976	1981	1986	Guinea-Bissau 1976	1981	1986
4.43[a]	5.53[a]	6.22[a]	0.66[a]	0.83[a]	0.91[a]
42.7	42.9	43.1	38.7	40.1	40.7
4.4/5.1	4.5/5.2	4.5/5.2	6.5/7.4	6.4/7.2	6.3/7.2
56/37[b]	...	...	61/2[b]	...	...
...	...	...	...	...	...
...	...	...	...	...	...
16.3	19.1	22.2	20.8	23.8	27.1
2.2	2.3	2.5	5.0	1.9	2.1
37/40	39/42	41/44	39/43	41/45	43/47
171	159	147	154	143	132
...	...	...	...	...	...
...	...	...	...	...	...
...	...	...	...	...	...
...	16	15[e]	...	16	21[e]
...	47	45[e]	...	51	51[e]
...	20	21[e]	...	5	4[e]
...	3	3[e]	...	2	2[e]
95	104	108	101	112	148
104	102	94	135	109	131
...	...	...	...	...	...
7[b]	7	14	...	...	...
165[b]	270	...	37	50	...
143[b]	390	...	6	14	...
...	...	...	...	...	...
...	...	...	...	...	...
21.25[h]	21.21[h]	235.63[h]	27.47[bk]	37.80[k]	238.98[k]
...	...	...	...	...	...
...	...	...	...	...	...
...	...	...	...	...	...
3	...	7[e]	...	...	...
...	2	...	...	...	...
-	1.2[d]	1.3[c]	...	...	...
...	...	...	...	67/91[f]	...
28/13[b]	32/15[d]	31/13[c]	54/23[b]	60/24	56/26[c]
...	...	3.3[e]	3.9	3.7	3.4[e]
17 510[g]	...	...	...	8 841[f]	...
1 934	1 768	1 724	...	...	...

United Nations estimate; **b:** 1975; **c:** 1985; **d:** 1980; **e:** 1984; **f:** 1979; **g:** 1977; **h:**
urrency: Sylis; **k:** Currency: Guinea-Bissau pesos.

Guyana

1976	1981	1986
0.79	0.88[a]	0.97[ab]
43.8	39.4	37.0
5.4/5.6	5.5/5.9	5.7/6.3
46/15	48/16[f]	...
29.1[e]	20.3[f]	...
23.9[e]	16.8[f]	...
29.6	30.5	32.2
2.1	2.0	1.9
64/69	66/71	67/72
49	36	30
504[c]	568[f]	462[d]
3.4	-0.9	-4.5
646[c]	683[f]	485[d]
34	31	23[p]
21	19	20[p]
25	19	15[p]
12	13	11[p]
99	104	91
108	102	81
...	...	...
0	0	0
364	436	...
279	346	218
59.7[h]	122.2[h]	208.8[hk]
...	...	...
2.55[t]	3.00[t]	4.40[t]
27	7	9
...	...	...
...	...	...
33[c]	52	...
27	34	42[d]
...	...	...
...	...	...
75/75[c]	79/80	...
4.2[c]	7.7	7.2[d]
7 604[cq]	9 965[qr]	...
2 343	2 412	2492

Haiti

1976	1981	1986
4.67	5.10	5.36
43.4	43.6	43.6
5.3/6.2	5.1/6.0	4.9/5.8
55/46[c]	51/34[g]	...
...	57.4[g]	...
...	6.7[g]	...
22.1	24.6	27.2
2.4	2.5	2.6
49/52	51/54	53/56
139	128	117
681[c]	1 479[f]	2 009
3.8	5.3	-0.9
132[c]	247[f]	305
16[cs]	17	12
41[c]	39	38
16[c]	16	15
14[c]	14	14
96	100	110
106	97	94
...	...	...
10[c]	19	27
201	448	...
118	151	170
...	111.0[n]	154.5
...	...	...
5.00[v]	5.00[v]	5.00
28	24	16
0.00	0.02	0.02
86	158	112
4	6	...
...	...	...
2.6[c]	2.8[f]	3
...	63/67[g]	...
...	45/39	54/47
1.0	1.2[f]	1.2
5 936[eq]	8 198[qr]	...
1 936	1 904	1 843

a: United Nations estimate; **b:** Marked break in series; **c:** 1975; **d:** 1985; **e:** 1977 1980; **g:** 1982; **h:** Urban areas; **k:** January-September; **n:** Metropolitan area; **p:** 198 **q:** Physicians in government service; **r:** 1979; **s:** Including increase in stocks; **t:** C rency: Guyana dollars; **v:** Currency: Gourdes.

Honduras 1976	1981	1986	Hungary 1976	1981	1986
3.20	3.82	4.51a	10.59	10.71	10.63
48.0	47.8	46.9	20.3	21.9	21.6
4.0/4.5	4.2/4.6	4.4/4.7	16.0/20.5	14.6/19.6	15.2/21.0
...	49/10d	...	56/41	55/40	51/41
...	52.9d	57.2e	21.9b	18.5	22.7c
...	12.1d	14.0e	34.7b	35.6	31.3c
32.3	36.1	40.0	50.1	53.5	56.2
3.5	3.4	3.1	0.3	-0.0	-0.1
55/59	58/62	61/64	67/73	67/74	68/75
95	82	69	27	20	17
1 121b	2 544d	3 480c	433pr	635pr	842cpr
2.6	7.3	0.7	6.6r	3.2r	1.4r
362b	689d	796c	40 887s	59 290s	79 061cs
21	19	17c	21	16	12c
27	21	19c	16	45	...
17	16	17c	15	37	...
4	13	13c	12	39	...
78	107	108	86	102	109
94	104	90	87	101	110
73	100	95c	89	103	112
36	69	76	14 361	15 066	15 130
453	945	890c	5 528	9 128	9 613
397	728	780c	4 932	8 712	9 183
67.4	109.4	139.1	77.4	104.6	146.5
...	...	...	...	...	...
2.00ft	2.00ft	2.00ft	20.83v	34.43v	45.93v
131	101	111	1 584b	1 652	3 062
0.00	0.02	0.02	0.95b	1.68	2.35
...	126	126e	5 551	10 450	10 613
6b	20	...	54b	116	156c
6b	8	11e	99b	121	140c
11b	13d	64c	218bg	258dg	272cg
...	...	39/42c	...	1/1d	...
60/59b	69/69d	75ch	87/84	89/88d	90/90c
...	3.4	...	3.5b	3.9d	6.4c
3 370b	3 124n	...	389b	390k	...
2 109	2 197	2 208	3 429	3 496	3 522

: United Nations estimate; **b:** 1975; **c:** 1985; **d:** 1980; **e:** 1984; **f:** Buying rate; **g:** Number of licences issued or sets declared; **h:** Both sexes; **k:** 1982; **n:** 1979; **p:** Billion Forint; **r:** Net material product; **s:** In national currency; **t:** Currency: Lempiras; **v:** Currency: Forint.

Iceland

1976	1981	1986
0.22	0.23	0.24
30.1	27.5	26.6
11.7/13.8	12.3/14.7	12.5/15.1
58/27	60/44	...
...	...	...
...	...	...
86.8	88.2	89.4
0.9	1.2	0.9
73/79	74/80	74/80
9	6	6
1 320[b]	3 232[h]	2 663[c]
6.0	6.6	0.2
6 055[b]	14 175[h]	10 959[c]
27	25	21[c]
9	10	...
20	22	...
17	19	...
92	102	97
96	100	91
...	...	...
201	277	352
470	980	1 115
404	859	1 096
23.1[d]	150.8[d]	870.6[d]
...	...	...
1.90[gp]	8.17[gp]	40.24[gp]
79	230	310
0.03	0.05	0.05
70	72	111
291[b]	396	488[c]
417[b]	480	...
234[bk]	281[hk]	305[ck]
...	...	...
91/87	...	97/96[c]
...	...	...
...	467	...
2 970	3 140	3 041

India

1976	1981	1986
613.27[a]	690.00[a]	766.14[a]
39.8	38.5	36.8
6.1/6.3	6.4/6.6	6.7/7.0
51/26[b]	53/20	...
...	62.6	...
...	11.1	...
21.5	23.4	25.5
2.1	1.9	1.7
53/52	56/55	58/58
126	110	99
88 758[b]	162 092[h]	196 904[c]
2.6	3.4	5.5
143[b]	235[h]	259[c]
19	20	22[c]
36	32	28[c]
17	18	19[c]
14	15	15[c]
91	106	123
99	104	111
88	109	149
63 356	92 177	139 576
5 665	15 654	14 870
5 548	8 373	9 207
75.7[e]	113.8[e]	148.0[e]
...	112[f]	150[f]
8.881[q]	9.099[q]	13.122[q]
2 792	4 693	6 396
6.95	8.59	10.45
534	1 279	1 451
1.3[b]	2.4	...
3[b]	5	5[c]
0.7[bk]	2.2[hk]	4.6[ck]
...	45/74	...
66/40[b]	69/44	72/47[n]
2.8	2.9	3.6[c]
3 039	2 545	...
1 932	2 104	2 161

a: Including the Indian-held part of Jammu and Kashmir, the final status of which ha not been determined; **b:** 1975; **c:** 1985; **d:** Reykjavik; **e:** Agricultural workers; excludin rent; **f:** Domestic supply; **g:** Selling rate; **h:** 1980; **k:** Number of licences issued (sets declared; **n:** 1984; **p:** Currency: Couronnes; **q:** Currency: Rupees.

Indonesia 1976	Indonesia 1981	Indonesia 1986	Iran (Islamic Republic of) 1976	Iran (Islamic Republic of) 1981	Iran (Islamic Republic of) 1986
133.53	149.70	166.94	33.71	39.54	45.91[a]
42.0	41.0	38.7	45.4	44.1	42.7
5.0/5.6	5.0/5.6	5.4/5.9	4.8/5.5	4.9/5.4	4.8/5.4
51/28[d]	48/23[e]	...	48/9	47/7	...
64.7	53.0[f]	...	36.9	...	...
6.7	10.9[f]	...	18.7	...	...
19.4	22.2	25.3	45.7	49.1	51.9
2.1	2.0	1.7	3.0	2.9	2.8
49/51	52/55	55/57	57/57	57/57	59/59
95	84	74	120	115	107
30 464[b]	72 482[d]	85 081[c]	51 924[b]	98 081[d]	168 100[c]
8.5	7.8	3.8	10.0	-5.0	6.7
225[b]	480[d]	511[c]	1 557[b]	2 539[d]	3 766[c]
21	21	20[c]	34	19	19[p]
31	25	24[c]	8	18	17[p]
29	35	31[c]	45	23	20[p]
9	11	14[c]	7	9	8[p]
81	107	133	93	110	117
88	105	119	103	107	98
56[g]	109[g]	141[g]	96[g]	107[g]	...
77 176	94 529	99 580	315 794	79 509[h]	103 238
5 673	13 272	10 718	12 894	12 549	...
8 547	25 164	14 805	23 503	12 587	12 378[c]
...	112.2	168.[illegible]	...	112.2	207.4[c]
51[k]	100	163[k]	...	...	...
415.0[r]	644.0[r]	1 641.0[r]	70.62[s]	79.45[s]	75.64[s]
1 497	5 014	4 051	8 681	1 605	...
0.06	3.10	3.10	3.74	6.06	...
401[n]	600[n]	825[n]	...	186	93
3[b]	9	12[c]	...	2	3
2[b]	4	...	20[b]	32[e]	54[c]
2.2[b]	20[e]	39[c]	51[b]	53[e]	56[c]
...	22/42[e]	...	52/76	...	...
62/49	81/69	86/77[p]	86/52	77/55	85/66[c]
2.1[b]	...	...	4.6[b]	6.4[e]	...
16 387[b]	11 681[q]	...	...	2 320[q]	...
2 058	2 440	2 504	...	...	...

United Nations estimate; **b:** 1975; **c:** 1985; **d:** 1978; **e:** 1980; **f:** 1982; **g:** Manufac-ing; **h:** Data relate to 12-month period ending 20 March; **k:** Excluding products of ectricity, gas and water; **n:** Visitor arrivals; **p:** 1984; **q:** 1979; **r:** Currency: Rupiahs; Currency: Iran rials.

Iraq			Ireland		
1976	1981	1986	1976	1981	1986
11.50	13.67	16.45[a]	3.23	3.44	3.54
46.6	47.0	46.9	31.2	30.6	29.6
3.8/4.4	3.9/4.5	4.0/4.6	14.3/16.7	13.6/16.1	13.0/15.8
42/9[d]	...	...	53/21[b]	53/21	52/22
30.1[d]	...	...	20.0[b]	15.4	14.1
11.0[d]	...	...	20.9	22.9	20.7
61.4	66.4	70.6	53.6	55.3	57.0
3.8	3.6	3.3	1.2	1.2	1.3
60/62	61/63	63/65	70/75	70/76	71/77
83	77	69	15	10	9
13 866[b]	53 643[n]	...	8 427[b]	19 261[n]	18 394
8.6	13.0	...	4.7	5.0	1.5
1 258[b]	4 036[n]	...	2 629[b]	5 663[n]	5 181
29	46	...	25	30	21
9	9	...	15	10	11
69	37	...	23	24	24
8	7	...	20	...	...
104	102	147	88	92	106
120	98	119	92	91	99
...	...	...	80[f]	105[f]	130
120 885	44 660	84 769	1 064	2 503	2 786
3 897	11 344	...	4 196	10 607	12 610
9 272	10 530	...	3 315	7 675	13 158
...	129.4[g]	190.9[g]	61.1	120.4	185.2
...	...	...	...	117[h]	152
0.295[p]	0.295[p]	0.311[p]	0.587[q]	0.633[q]	0.715
4 434	...	...	1 818	2 651	3 236
4.10	...	...	0.45	0.36	0.36
630[k]	1 564[k]	1 004[k]	1 690	2 188	2 378
10[b]	33	...	162[b]	242	229
17[b]	...	...	138[b]	208	265
37[b]	50[n]	57[c]	...	231[n]	252
...	...	10/12[c]	...	...	...
89/45[b]	100/77[n]	91/69[c]	94/98[b]	94/99	96/100
3.3	...	...	5.3[b]	5.7[n]	6.1
2 587[b]	1 792	...	830[b]	774	...
...	...	...	3 492	3 713	3 795

a: United Nations estimate; **b:** 1975; **c:** 1985; **d:** 1977; **e:** 1984; **f:** Excludes electric gas and water; **g:** Base: 1979 = 100; **h:** Domestic supply; including exported produc **k:** Visitor arrivals; **n:** 1980; **p:** Currency: Iraqi dinars; **q:** Currency: Irish pounds.

Israel 1976	Israel 1981	Israel 1986	Italy 1976	Italy 1981	Italy 1986
3.53	3.95	4.30	55.70	56.51	57.22
32.8	32.6	31.7	24.0	22.0	19.4
11.4/12.2	11.9/13.2	11.3/13.4	15.4/19.3	15.0/19.4	16.3/20.4
44/22	43/25[c]	43/26[b]	52/20	54/27	55/28[b]
6.3	6.1[c]	4.2[b]	15.1	9.9	9.8[b]
25.0	23.9[c]	23.0[b]	32.3	27.3	21.3[b]
86.6	88.6	90.3	65.6	66.5	67.4
2.3	1.8	1.7	0.4	0.1	0.1
71/75	73/76	73/77	70/77	71/78	72/79
18	14	14	18	13	11
12 520[a]	22 443[c]	24 559[b]	192 047[a]	395 520[c]	358 669[b]
7.7	3.3	2.2	3.1	3.6	0.8
3 624[a]	5 787[c]	5 776[b]	3 440[a]	6 930[c]	6 259[b]
27	22	18[b]	20	20	18[b]
9	5	4[b]	7	6	5[b]
18	17	18[b]	36	33	31[b]
...	...	...	...	...	...
97	100	110	89	102	101
114	94	103	90	101	100
87	106	...	86	98	99
84	148	49	19 633[e]	18 509[e]	20 284[e]
4 077[t]	7 847[t]	9 347[t]	43 933	91 022	99 937
2 306[t]	5 329[t]	6 846[t]	37 281	75 187	97 835
12	217	33 322	54.1	117.8	201.5
...	208[f]	776[fg]	...	112[h]	171[h]
...	0.016[r]	1.49[r]	875.0[s]	1 200.0[s]	1 358.1[s]
1 328	3 497	4 660	3 300	20 134	19 987
1.10	1.19	1.02	82.48	66.67	66.67
733[k]	1 090[k]	1 101[k]	13 930[k]	20 036[k]	24 672[k]
82[a]	140	172[b]	272[a]	348	405[b]
219[a]	313	385[b]	259[a]	364	448[b]
...	232[c]	259[b]	217[an]	234[cn]	253[dn]
...	5/11[g]	...	...	2/4	2/4[b]
85/89[a]	88/91[p]	90/94[b]	87/82	83/82	...
5.7[a]	7.2	...	4.1[a]	5.2[g]	...
...	376[p]	...	...	346[q]	...
3 063	2 993	3 049	3 461	3 622	3 486

a: 1975; **b:** 1985; **c:** 1980; **d:** 1985; **e:** Including data for San Marino; **f:** Agricultural products; **g:** 1983; **h:** Domestic supply; including exported products; excluding products of electricity, gas and water; **k:** Excluding nationals residing abroad; **n:** Number of licences issued or sets declared; **p:** 1982; **q:** Including dentists; **r:** Currency: new shekels; **s:** Currency: Lires; **t:** Excluding Judea and Samaria and the Gaza area.

Jamaica

1976	1981	1986
2.08	2.16	2.37[a]
45.2	40.6	36.7
8.0/9.0	7.9/9.1	7.8/9.3
47/38	50/43[g]	...
29.0[e]	26.5[g]	...
14.5[f]	15.6[dg]	...
45.7	49.8	53.8
1.2	1.4	1.5
67/71	70/76	71/77
25	21	18
2 861[b]	2 667[h]	2 026[c]
1.9	-2.6	0.4
1 400[b]	1 227[h]	867[c]
17	18	22[c]
8	8	6[c]
29	28	28[c]
18	16	20[c]
96	96	108
101	94	99
...	...	...
9	11	13
913	1 473	981
630	974	596
40.6	112.1	247.8
...	...	...
0.909[q]	1.78[q]	5.48[q]
32	85	98
...	0.00	0.00
328	406	663
29[b]	57	...
50[b]	62	...
54[b]	77[h]	92[c]
...	...	...
75/81[b]	76/81[h]	...
4.7[b]	6.8[h]	5.4[c]
3 509[b]	3 031[h]	...
2 669	2 572	2 576

Japan

1976	1981	1986
112.77	117.65	121.67
24.3	23.6	21.8
10.5/12.9	11.2/14.5	12.4/16.5
62/35[b]	62/36[h]	60/39[c]
13.6[b]	10.7[h]	8.5[c]
25.1[b]	23.9[h]	25.1[c]
75.7[b]	76.2	76.5[c]
1.4	0.9	0.6
72/77[b]	73/79[h]	75/81
10	7	5
499 774[b]	1 059 262[h]	1 325 203[c]
4.6	5.0	3.8
4 481[b]	9 068[h]	10 975[c]
31	31	28[c]
5	3	3[c]
33	32	34[c]
31	29	30[c]
104	98	109
106	97	106
80	101	121
25 574	28 729	34 532
64 797	143 288	126 408
67 224	152 016	209 153
79.7	104.8	115.2
...	101[k]	90[k]
292.8[r]	219.9[r]	159.1[r]
15 746	28 208	42 257
21.11	24.23	24.23
795	1 069	1 842
155[b]	336	374[c]
354[b]	460[h]	555[c]
...	539[h]	580[c]
...	...	...
95/96[b]	97/98[h]	98/99[c]
3.9[b]	4.0	...
845[b]	780[n]	735[p]
2 782	2 851	2 804

a: United Nations estimate; **b:** 1975; **c:** 1985; **d:** Including construction and sanita services; **e:** Including mining and quarrying; **f:** Including transportation, storage, com munication and sanitary services; **g:** 1982; **h:** 1980; **k:** Domestic supply; including e ported products; **n:** 1979; **p:** 1984; **q:** Currency: Jamaican dollars; **r:** Currency: Ye

ordan			Kenya		
1976	1981	1986	1976	1981	1986
2.78	3.02[a]	3.66[a]	13.85	17.34	21.16
47.2	49.4	48.2	57.7	52.2	52.5
4.1/4.6	4.5/4.7	4.0/4.4	2.8/3.5	2.7/3.3	2.7/3.2
36/4[b]	38/3[d]	...	52/26[b]	...	...
...	10.3[d]	...	...	...	...
...	9.1[d]	...	...	...	...
55.4	60.1	64.4	12.9	16.1	19.7
2.3	3.7	4.0	4.0	4.1	4.2
59/63	62/66	64/68	49/52	51/55	53/57
65	54	44	88	80	72
1 007[b]	3 303[h]	4 067[c]	3 253[b]	7 088[h]	5 769[c]
0.1	11.7	3.9	4.5	6.6	2.1
387[b]	1 130[h]	1 157[c]	237[b]	423[h]	280[c]
32	49	27[c]	20	24	18[c]
9	6	7[c]	33	28	27[c]
13	19	19[c]	12	13	13[c]
8	14	14[c]	10	11	11[c]
72	111	124	94	100	127
81	107	100	117	94	92
...	...	...	68[e]	105[e]	...
...	...	15	68	120	174
1 022	3 149	2 432	973	2 069	1 613
209	732	647	825	1 188	1 200
64.3	107.7	130.0	68.0[f]	113.8[f]	215.2[f]
...	...	...	...	...	...
0.331[p]	0.339[p]	0.344[p]	8.310[q]	10.286[q]	16.042[q]
472	1 087	438	276	231	413
0.80	1.07	1.06	0.00	0.08	0.08
1 063[g]	1 581[g]	1 912[g]	446	373	604
12[b]	41	59[c]	7[b]	14	11[c]
...	...	...	9[b]	13	13[k]
46[b]	59[h]	68[c]	2.8[b]	4.0[h]	4.9[c]
19/46	...	...	...	...	...
...	93/90[h]	...	73/60[b]	79/70[h]	75/67[c]
4.3[b]	5.2[h]	4.4[c]	6.0[b]	6.4[h]	6.3[c]
3 632[b]	1 704[h]	1 309[k]	10 134[n]	...	...
...	...	...	2 249	2 192	2 162

United Nations estimate; **b:** 1975; **c:** 1985; **d:** 1979; **e:** Manufacturing; **f:** Nairobi; iddle-income group; **g:** Visitor arrivals; **h:** 1980; **k:** 1984; **n:** 1978; **p:** Currency: Jor-anian dinars; **q:** Currency: Kenyan shillings;.

Kiribati | Kuwait

Kiribati 1976	Kiribati 1981	Kiribati 1986	Kuwait 1976	Kuwait 1981	Kuwait 1986
0.05[a]	0.06[a]	0.06[a]	1.07	1.43	1.79
...	...	...	44.4	40.2	40.0
...	...	...	2.4/2.8	2.1/2.5	2.4/2.5
...	...	...	50/8[b]	55/11[d]	...
...	...	...	2.5[b]	1.9[d]	...
...	...	...	12.0[b]	11.4[d]	...
...	...	...	83.8	90.2	93.5
...	...	...	6.2	5.5	4.2
...	...	...	67/72	70/74	70/75
...	...	...	34	23	20
...	...	...	12 017[b]	28 726[d]	19 744[c]
...	...	...	-4.8	1.6	-3.8
...	...	...	11 993[b]	20 892[d]	10 902[c]
...	...	...	15	16	15[c]
...	...	...	0	0	0[c]
...	...	...	72	62	53[c]
...	...	...	6	5	6[c]
...	...	...	...	...	...
...	...	...	...	...	...
...	...	...	...	...	...
...	...	...	112 995[e]	64 534[e]	79 759[e]
...	...	...	3 327	6 969	...
...	...	...	9 846	16 298	10 126[c]
65.2	107.3[f]	130.9[cf]	74.8	107.3	125.6
...	...	...	...	107[g]	...
0.795[bp]	...	1.504[p]	0.287[q]	0.281[q]	0.292[q]
...	...	...	1 702	4 068	5 501
...	...	...	5.58	2.54	2.54
...	...	...	...	128[h]	84[h]
...	...	...	202[b]	394	314[c]
...	...	...	130[b]	159	...
...	...	...	...	258[d]	235[c]
...	...	...	32/52[b]	27/41[d]	...
...	...	...	83/72[b]	92/85[d]	91/87[c]
4.8[b]	12.3[d]	8.7[k]	2.4[b]	2.3[d]	4.5[c]
2 077[b]	1 933	...	914[b]	635[n]	...
2 456	2 672	2 616	2 794	3 162	3 135

a: United Nations estimate; **b:** 1975; **c:** 1985; **d:** 1980; **e:** Part of Neutral Zone; Tarawa; **g:** Domestic supply; **h:** Arrivals of tourists at hotels; **k:** 1984; **n:** 1982; **p:** Cu rency: Kiribati dollars; **q:** Currency: Kuwaiti dinars.

Lao People's Democratic Republic | Lebanon

Lao 1976	Lao 1981	Lao 1986	Lebanon 1976	Lebanon 1981	Lebanon 1986
3.52[a]	3.75[a]	4.22[a]	2.77[a]	2.65[a]	2.71[a]
42.9	43.2	42.5	41.2	40.1	37.5
4.5/5.0	4.6/5.2	4.7/5.4	7.1/8.0	7.0/7.7	7.5/8.0
54/44[b]	...	...	42/10[b]	...	...
...	...	...	17.0[b]	...	...
...	...	...	19.2[b]	...	...
11.4	13.4	15.9	68.6	74.8	80.1
1.4	2.2	2.4	-0.72	-0.01	-2.13
46/49	48/51	50/53	63/67	63/67	65/69
135	122	110	48	48	39
...	...	...	...	4 075	...
...	...	...	...	...	...
...	...	...	...	...	...
...	...	...	...	1 527	...
...	...	...	...	...	...
...	...	...	...	...	...
...	...	...	...	...	...
67	111	146	62	94	128
70	109	129	58	94	127
...	...	...	...	...	...
29[b]	93	86	68	72	48
42[b]	125	48[d]	838	3 615	...
31[b]	33	12[d]	496	886	...
...	...	...	...	...	...
...	...	...	...	...	...
0.375[bk]	30.00[k]	95.00[k]	2.930[p]	4.625[p]	87.00[p]
...	...	...	1 303	1 516	488
...	...	...	9.21	9.22	9.22
...	29	...	...	...	...
...	...	...	76[b]	130	...
2[b]	...	...	...	...	...
...	...	...	148[b]	281[e]	300[c]
...	...	8/24[cf]	...	...	...
...	60/49	61/47[d]	...	83/81	77/72[d]
...	...	...	...	...	...
21 667	...	...	...	260[h]	...
...	...	...	...	...	...

United Nations estimate; **b:** 1975; **c:** 1985; **d:** 1984; **e:** 1980; **f:** Age group: 15-45 ears; **h:** 1979; **k:** Currency: new kip; **p:** Currency: Lebanese pounds.

Lesotho 1976	Lesotho 1981	Lesotho 1986	Liberia 1976	Liberia 1981	Liberia 1986
1.21	1.37	1.56[a]	1.61	1.91	2.22
41.7	42.0	42.3	44.1	45.6	46.8
4.8/6.5	4.8/6.5	4.9/6.4	4.8/5.3	4.7/5.2	4.6/5.1
61/48	...	...	...	...	...
...	...	...	...	...	...
...	...	...	...	...	...
10.8	13.6	16.7	30.4	34.9	39.5
2.4	2.5	2.6	3.4	3.2	3.2
44/50	46/52	48/54	45/49	47/51	49/53
123	111	100	143	132	122
152[b]	422[k]	...	610[b]	917[k]	811[c]
9.1	9.5	...	2.7	2.5	-2.9
128[b]	315[k]	...	386[b]	490[k]	370[c]
27	31	...	33	20	15[c]
35	23	...	12	13	17[c]
5	10	...	38	26	26[c]
4	5	...	7	8	7[c]
88	98	94	89	102	116
98	96	79	104	101	97
...	...	...	...	...	...
d	d	d	28	27	27
d	d	d	399	477	235
d	d	d	457	529	404
55.5[f]	116.5[f]	229.4[f]	69.2[g]	108.6[g]	123.2[g]
...	...	...	...	...	...
0.870[p]	0.957[p]	2.183[p]	1.00[hq]	1.00[hq]	1.00[h]
...	43	60	17	8	3
...	...	...	...	...	...
...	55	132	...	...	...
6[b]	15	...	6[b]	10	9[c]
...	...	10[c]	...	...	...
...	...	0.3[c]	5.7[b]	11[k]	16[c]
...	...	...	...	...	...
59/82[b]	59/82[k]	69/88[n]	56/28[b]	67/37[k]	...
3.2[b]	3.7[k]	2.8[n]	...	4.9[k]	...
24 330[b]	...	...	9 141[b]	...	...
2 080	2 347	2 346	2 281	2 381	2 342

a: United Nations estimate; **b:** 1975; **c:** 1985; **d:** Data included with South Africa; 1982; **f:** Low income group; **g:** Monrovia; **h:** Buying rate; **k:** 1980; **n:** 1984; **p:** Currency: Maloti; **q:** Currency: Liberian dollars.

ibyan Arab amahiriya 1976	1981	1986	Liechtenstein 1976	1981	1986
2.55	3.10	3.74[a]	0.02	0.03	0.03
46.1	46.7	46.4	24.7	22.3	20.1
3.6/3.9	3.6/3.9	3.8/3.9	11/14	11/14	12/16
...	...	...	62/25	64/31	63/33
...	...	...	4.2	3.4	2.8
...	...	...	54.6	52.5	43.8
46.8	56.6	64.5	-	-	-
4.0	3.9	3.7	1.9	1.6	1.2
54/57	57/60	59/62	63/71	64/69	67/75
107	97	82	12	9	7
12 770[b]	35 592[e]	...	252	485	782
7.8	8.1	...	...	...	...
5 255[b]	11 972[e]	...	9 640	16 582	21 788
25	30	...	...	...	...
2	2	...	...	...	...
59	54	...	...	...	...
2	3	...	...	...	...
108	96	177	...	...	...
127	92	141	...	...	...
...	...	...	67	99	145
98 957	62 636	54 439	d	d	d
3 212	8 382	...	...	...	...
9 561	14 371	10 841[c]	239	450	723
...	...	...	91	106	124
...	...	...	95	106	111
0.296[bg]	0.296[g]	0.314[g]	2.50[h]	1.96[h]	1.79[h]
3 106	9 003	5 953	...	...	...
2.44	3.58	3.60	...	...	...
132	126	...	81	91	83
88	242	...	410	524	534
...	...	...	672	820	968
35[b]	55[e]	65[c]	189	279	317
...	...	...	...	...	...
...	...	...	...	...	...
3.7[b]	2.3[e]	...	3.3	3.5	3.5
940[b]	619[f]	...	711	670	571
3 468	3 653	3 619	...	...	...

Jnited Nations estimate; **b:** 1975; **c:** 1985; **d:** Included with data for Switzerland; 1980; **f:** 1982; **g:** Currency: Libyan dinars; **h:** Currency: Swiss francs.

Luxembourg			Madagascar		
1976	1981	1986	1976	1981	1986
0.36	0.37	0.36[a]	7.81[a]	8.96[a]	10.30
21.6	20.6	18.4	42.6	43.4	44.2
16.1/21.1	14.8/20.2	14.6/20.9	5.1/5.7	5.1/5.8	5.2/5.8
58/20[c]	58/21	60/29[d]	56/45[c]	...	...
...	4.9	...	...	...	...
...	23.3	...	...	...	...
73.7	77.6	81.0	16.3	18.9	21.8
0.1	-0.0	-0.1	2.7	2.8	2.9
67/74	68/74	69/75	47/48	49/50	51/52
13	9	8	75	67	59
2 352[c]	4 544[h]	3 567[b]	1 844[c]	3 265[h]	2 345
4.2	2.6	2.7	0.2	1.7	-0.7
6 497[c]	12 484[h]	9 826[b]	243[c]	375[h]	234
25	25	20	12[n]	19	...
3	3	3[d]	40	40	42
35	30	33[d]	18	16	16
32	27	30[d]	...	...	...
e	e	e	100	104	113
e	e	e	112	101	96
97	94	124	...	...	...
44	13	7	14	13	23
e	e	e	285	540	402
e	e	e	275	316	274
81.8[g]	108.1[g]	140.2[g]	67.5[f]	130.5[f]	285.5
...	...	...	...	...	...
35.98[p]	38.46[p]	40.41[p]	248.50[q]	287.40[q]	769.810
...	...	...	42	37	115
...	0.46	0.43	...	...	...
...	...	622[b]	10	12	27
319[c]	464	485[b]	7[c]	12	...
242[c]	366[h]	...	4[c]	4[h]	...
238[c]	247[h]	252[b]	1.1[c]	5.2[h]	5
...	...	...	...	...	...
...	82/80[k]	...	69/51[c]	...	82/71
3.8[c]	5.7	...	3.0[c]	4.6[h]	3.3
890[c]	737	...	10 665[c]	9 939	...
e	e	e	2 533	2 510	2 467

a: United Nations estimate; **b:** 1985; **c:** 1975; **d:** 1984; **e:** Data included with Belgi
f: Antananarivo, excluding rent; **g:** Excluding rent; **h:** 1980; **k:** 1983; **n:** 1977; **p:** (
rency: Luxembourg francs; **q:** Currency: Malagasy francs.

Malawi 1976	Malawi 1981	Malawi 1986	Malaysia 1976	Malaysia 1981	Malaysia 1986
5.37	6.23	7.28	12.24	14.10a	16.11
47.2	46.5	46.0	42.3	39.3	37.8
3.4/4.0	3.7/4.4	3.8/4.6	5.6/5.4	5.6/5.8	5.4/6.0
46/37d	...	52/52e	...	50/25f	...
84.4d	...	...	...	37.7f	...
3.9d	...	...	...	13.2f	...
7.7	9.7	12.0	30.5	34.2	38.2
2.8	3.1	3.3	2.2	2.4	2.1
42/44	44/46	46/48	63/67	65/69	66/71
177	163	150	34	30	26
613b	1 225f	1 203q	9 297b	24 487f	31 231c
8.7	6.9	2.3	10.3	8.3	5.5
119b	206f	179q	755b	1 779f	2 008c
22	16	13q	22	36	30c
38	...	...	25s	...	...
14	...	...	31	...	...
13	...	...	19s	...	...
88	102	112	89	103	122
101	100	88	92	102	112
80g	110g	120g	73	103e	151
24	34	44	8 690	12 731	33 979
206	350	252	3825	11 546	10 839
166	270	243	5 293	11 766.	13 869
67.2h	109.6h	210.7h	82.3k	109.7k	126.4k
...	...	...	...	...	...
0.907t	0.907t	1.952t	2.535v	2.242v	2.603v
26	49	25	2 404	4 098	6 027
...	0.01	0.01	1.66	2.33	2.34
50	49	54c	1 225n	2 345n	3 027n
2.1b	4.5	4.1c	40b	68	117c
4b	5f	6c	25b	53	81c
...	...	...	36bp	81fp	101cp
...	...	...	...	20/40f	...
52/32b	54/36f	52/38q	70/63b	71/68f	75/75c
2.2b	2.5f	2.6c	5.1b	5.0f	5.7c
50 040	40 950r	...	2 744k	...	...
2 478	2 472	2 429	2 492	2 596	2 634

United Nations estimate; **b:** 1975; **c:** 1985; **d:** 1977; **e:** 1983; **f:** 1980; **g:** Manufac-ring; **h:** Blantyre; low income group; **k:** Peninsular Malaysia component; **n:** Foreign urist departures; **p:** Number of licences issued or sets declared; **q:** 1984; **r:** 1979; 1978; **t:** Currency: Kwacha; **v:** Currency: Ringgit.

Maldives			Mali		
1976	1981	1986	1976	1981	1986
0.13	0.16	0.19	6.32	7.29	8.44
...	...	...	45.4	46.0	46.3
...	...	...	4.1/4.9	4.1/4.9	4.1/4.9
...	...	...	60/12	...	...
...	...	...	82.2	...	...
...	...	...	1.2	...	...
18.1[a]	23.6	...	16.2	17.3	18.0
1.8	3.2	3.2	2.2	2.8	2.9
...	...	...	39/42	40/44	42/46
...	...	...	191	180	169
27[a]	47[c]	84[b]	743[a]	842[c]	538[c]
13.5	13.5	6.6	1.9	4.7	0.1
203[a]	303[c]	459[b]	118[a]	120[c]	69[c]
...	...	39[d]	14	16[c]	...
...	35	29[d]	...	46	...
...	5	6[d]	...	7	...
...	4	5[d]	...	...	...
...	...	...	87	106	123
...	...	...	96	104	103
...	...	...	...	...	...
...	...	...	3[a]	4	11
13[a]	17[c]	64[d]	299	385	438
4[a]	8[c]	14[d]	172	154	192
...	...	...	...	...	...
...	...	...	...	...	...
8.625[h]	7.550[h]	7.244[h]	...	287.40[k]	322.75[k]
...	...	...	7	17	23
...	...	...	...	0.02	0.02
...	74	114	20	28	...
...	...	...	1.9	4.1	...
...	...	...	...	...	...
-	7[c]	17[b]	-	-	0.1[b]
18/18[e]	...	...	86/94	...	...
...	...	...	23/11[a]	22/12	...
0.7[a]	...	...	3.4[a]	3.7	3.2[c]
15 555[e]	...[n]	...	22 850[f]	...	...
...	...	...	1 852	1 752	1 793

a: 1975; **b:** 1985; **c:** 1980; **d:** 1984; **e:** 1977; **f:** 1978; **h:** Currency: Rufiyaa; **k:** Curre
cy: CFA francs.

alta			Mauritania		
1976	1981	1986	1976	1981	1986
0.33	0.36	0.38[a]	1.46[a]	1.68[a]	1.95[a]
24.7	23.0	23.9	45.3	45.7	46.4
.5/13.7	12.8/14.5	12.6/15.0	4.3/5.0	4.3/5.0	4.2/4.9
57/19[b]	59/20[d]	57/17[e]	60/3[b]	...	...
6.4[b]	5.8[d]	...	...	...	...
33.1[b]	34.0[d]	...	...	...	...
80.7	83.1	85.3	19.6	26.9	34.6
1.3	0.7	0.7	2.8	2.9	3.1
69/73	69/74	70/75	40/44	42/46	44/48
15	13	11	149	137	127
434[c]	1 136[d]	1 017[e]	478[c]	829[d]	697[k]
9.2	11.5	1.4	3.2	1.1	-1.9
1 254[c]	3 079[d]	2 655[e]	336[c]	508[d]	380[k]
26	24	26[e]	45	22	33[k]
5	3	4[e]	25	35	19[k]
36	37	36[e]	23	17	19[k]
30	28	27[e]	4	5	9[k]
91	101	119	85	105	110
96	100	114	95	102	92
64	102	...	...	...	...
...	...	...	...	...	...
423	855	887	179	265	221
228	448	497	178	261	349
70.0	111.5	101.0	...	104.9	...
...	...	...	...	...	...
0.426[q]	0.387[q]	0.369[q]	43.64[r]	48.94[r]	74.08[r]
608	1 074	1 145	82	162	48
0.35	0.46	0.47	...	0.01	0.01
339[f]	706[f]	518[f]	...	...	...
164	234	264[e]	...	...	...
163	285	354[e]	...	3[h]	...
169[cg]	205[dg]	304[eg]	-	-	0.3[e]
...	...	...	83[n]	...	...
88/79	86/77[d]	90/83[k]	...	31/15[d]	...
3.7[c]	3.0[d]	3.3[e]	3.8	5.9	7.4[h]
859[c]	872[p]	...	15 172[c]	...	...
2 874	2 735	2 590	1 799	1 998	2 076

United Nations estimate; **b:** 1977; **c:** 1975; **d:** 1980; **e:** 1985; **f:** Departures; **g:** mber of licences issued or sets declared; **h:** 1983; **k:** 1984; **n:** Both sexes; **p:** 1982; Currency: Lire; **r:** Currency: Ouguiya.

Mauritius | Mexico

Mauritius 1976	Mauritius 1981	Mauritius 1986	Mexico 1976	Mexico 1981	Mexico 1986
0.89[a]	0.94[a]	0.99[a]	61.80	71.30	79.56
39.7	34.1	31.6	46.3	44.7	42.2
3.6/5.4	4.1/6.0	4.8/6.6	4.9/5.7	4.7/5.6	4.9/5.8
53/14[c]	...	56/19[b]	43/12	48/18[d]	...
28.2[c]	...	...	40.9[c]	25.8[d]	...
12.7[c]	...	...	19.6[c]	14.4[d]	...
43.6	42.9	42.2	62.8	66.4	69.6
1.9	1.9	1.7	2.9	2.6	2.4
63/67	64/69	66/71	62/66	64/68	65/70
38	28	23	60	53	47
661[c]	1 132[d]	1 061[h]	88 004[c]	186 331[d]	177 475
9.3	3.9	4.2	6.9	6.8	0.8
762[c]	1 185[d]	1 010[h]	1 463[c]	2 685[d]	2 247
27	22	19[b]	21	26	18
20	12	13[h]	11	8	9
15	15	19[h]	27	30	35
13	13	17[h]	23	22	24
112	101	121	82	106	111
124	98	107	91	104	97
...	...	...	72[e]	109[e]	101
5[c]	5	9	61 203	159 751	164 013
359	554	675	6 032	24 161	11 996
265	324	675	3 431	20 041	15 698
51.9	114.5	123.2[f]	45	128	1 995
...	...	...	...	...	...
6.639[k]	10.329[k]	13.137[k]	19.950[p]	26.229[p]	923.500
90	35	136	1 188	4 074	5 670
...	0.04	0.04	1.60	2.26	2.57
93	131	165	3 107	4 031	4 625
21[c]	42	44[b]	40[c]	95	91
29[c]	44	...	49[c]	78	...
46[cg]	87[dg]	102[gh]	...	55[d]	108
...	...	...	...	14/20[d]	8/12
72/68[c]	74/72[d]	78/76[h]	80/71[c]	89/85[d]	88/86
3.1[c]	4.7[d]	3.5[h]	2.5[c]	3.2	2.4
2 630	1 903[d]	...	...	...	...
2 556	2 723	2 721	2 713	3 053	3 147

a: Excluding Rodrigues and other small islands; **b:** 1984; **c:** 1975; **d:** 1980; **e:** Incluc construction; **f:** 1982 = 100; **g:** Number of licences issued or sets declared; **h:** 19 **k:** Currency: Mauritian Rupees; **p:** Currency: Mexican pesos.

Mongolia 1976	Mongolia 1981	Mongolia 1986	Morocco 1976	Morocco 1981	Morocco 1986
1.47	1.71	1.94	17.83	20.65	22.48[a]
43.8	43.1	41.6	47.2	43.2	41.0
4.7/5.4	4.7/5.4	4.9/5.6	5.3/5.1	6.4/6.0	5.8/5.5
50/26	...	...	44/8[b]	48/12[c]	...
...	...	...	...	39.2[c]	...
...	...	...	...	16.9[c]	...
48.7	51.1	50.8	37.8	41.3	44.8
2.8	2.7	2.7	2.4	2.3	2.5
59/63	60/64	62/67	54/58	57/60	59/62
62	53	45	110	97	82
...	...	...	8 983[b]	17 821[e]	11 892[d]
...	...	...	4.9	6.8	3.2
...	...	...	519[b]	919[e]	542[d]
...	...	...	29	22	21[d]
...	...	...	19	15	18[d]
...	...	...	24	27	25[d]
...	...	...	17	17	17[d]
99	101	110	98	88	137
111	98	96	107	86	118
72	110	...	81	100	...
757	1 136	1 620	654	677	706
301	655	...	2 617	4 356	4 069
232	436	...	1 262	2 320	2 640
...	...	...	68.3	112.5	174.4
...	...	...	...	...	...
3.33[h]	3.15[h]	3.00[h]	4.484[k]	5.333[k]	8.712[k]
...	...	...	467	230	211
...	...	...	0.61	0.70	0.70
...	...	...	1 108	1 567	2 128
...	...	...	19[b]	28	33[d]
21[b]	...	...	10[b]	12	14[d]
2.4[b]	3.0[e]	31[d]	26[bf]	39[ef]	52[df]
...	...	...	...	...	...
89/91[b]	91/96	91/96[g]	47/27[b]	62/38[e]	63/41[d]
...	...	...	4.4[b]	5.5	...
488	440	...	13 860	16 355	...
2 510	2 715	2 811	2 572	2 728	2 688

United Nations estimate; **b:** 1975; **c:** 1982; **d:** 1985; **e:** 1980; **f:** Number of licences sued or sets declared; **g:** 1984; **h:** Currency: Tughrik; **k:** Currency: Dirhams.

Mozambique 1976	Mozambique 1981	Mozambique 1986	Nauru 1976	Nauru 1981	Nauru 1986
10.16[a]	12.45	14.17	7[ak]	7[ak]	8
43.5	43.2	43.2	...	...	...
4.9/5.8	4.9/5.7	4.9/5.8	...	...	...
60/20	...	...	...	...	...
...	83.8	...	...	...	...
...	6.2[b]	...	...	...	...
8.6	13.1	19.4	...	...	...
4.4	2.8	2.7	1.2	0.6	3.1
43/45	44/46	46/48	...	...	...
165	153	141	...	...	...
...	...	...	...	...	...
...	...	...	...	...	...
...	...	...	...	...	...
...	...	...	...	...	...
...	...	...	...	...	...
...	...	...	...	...	...
...	...	...	...	...	...
97	102	100	...	...	...
115	98	85	...	...	...
...	...	...	...	...	...
508	524	8	...	...	...
288	801	482[c]	22[e]	32[d]	33
144	281	86[c]	82[e]	125[d]	122
...	102[g]	261[fg]	...	...	...
...	...	...	...	...	...
27.47[ehp]	33.80[hp]	39.36[hp]	0.795[eq]	0.847[dg]	1.208
...	...	...	...	...	...
...	...	...	...	...	...
...	...	...	...	...	...
...	11.4	3[f]	...	...	...
5[e]	4	4[f]	...	...	...
0.1[e]	0.1[d]	0.5[f]	...	...	...
...	...	...	...	...	...
57/38	48/35	44/32[f]	...	...	...
...	...	...	...	...	...
37 323[e]	33 883[d]	...	...	...	...
1 937	1 805	1 664	...	...	...

a: United Nations estimate; **b:** Excluding electricity, gas and water; **c:** 1984; **d:** 19
e: 1975; **f:** 1985; **g:** Maputo; **h:** Operational exchange rate for United Nations p
grammes as of 1 October; **k:** In thousands; **p:** Currency: Meticais; **q:** Currency: Nau
dollars.

epal 1976	epal 1981	epal 1986	Netherlands 1976	Netherlands 1981	Netherlands 1986
12.84	15.02	17.13	13.77	14.25	14.56
42.9	43.5	43.3	25.3	22.3	19.6
4.9/5.2	4.7/4.7	4.8/5.1	13.2/17.2	13.5/17.9	14.3/18.7
58/39[b]	...	...	54/21[c]	53/24	...
89.9[b]	91.1	...	5.6[c]	5.0	4.5[d]
0.7[b]	0.6	...	23.0[c]	19.7	17.8[d]
4.8	6.1	7.7	88.4	88.4	88.4
2.4	2.3	2.3	0.5	0.5	0.3
45/43	47/45	49/47	72/79	73/80	73/80
147	139	128	10	8	7
1 506[b]	1 946[g]	2 288[d]	86 975[b]	169 386[g]	124 983[d]
2.0	2.6	3.4	3.4	2.6	0.7
116[b]	133[g]	139[d]	6 305[b]	11 971[g]	8 620[d]
14	16	20[d]	19	19	19[d]
66	57	58[d]	4	4	4[d]
4	4	5[d]	28	26	29[d]
4	4	4[d]	22	17	18[d]
98	105	111	87	108	111
108	103	97	89	107	108
...	...	...	97	98	106
11	18	36	81 954	81 081	66 297
163	390	464	39 576	65 921	75 580
97	125	141	40 215	68 732	80 555
70.7	111.1	184.6	81.3	106.7	122.9
...	...	...	...	109[f]	107[f]
...	13.20[k]	...	2.457[p]	2.468[p]	2.192[p]
128	202	87	5 178	9 339	11 191
0.13	0.15	0.15	54.33	43.94	43.94
92	162	223	2 846	2 846	3 142
0.8[b]	...	...	249[b]	351	355[h]
1[b]	...	...	368[b]	539	609[d]
-	-	1.2[d]	...	399[g]	462[d]
66/95[b]	68/91	...	...	...	...
44/8	62/22	72/30[h]	95/91	97/96	99/98[h]
...	...	...	6.8[b]	6.9[g]	6.1[h]
36 453[b]	28 768[g]	...	630	510[e]	...
1 881	1 974	2 048	3 448	3 352	3 355

Marked break in series; **b:** 1975; **c:** 1977; **d:** 1985; **e:** 1982; **f:** Agricultural products 'op-growing production only); excluding livestock production, forestry, fishing and nting; **g:** 1980; **h:** 1984; **k:** Currency: Nepalese rupees; **p:** Currency: Guilders.

New Zealand

1976	1981	1986
3.11	3.13	3.25
30.0	26.7	24.1
11.3/14.2	12.3/15.7	12.9/16.3
55/26	56/29	...
12.0	10.8	...
25.2	24.9	...
82.8	83.3	83.7
0.5	0.9	0.9
69/76	71/77	71/78
14	12	11
14 177[f]	22 389[c]	22 367[g]
4.6	0.0	3.1
4 592[f]	7 065[c]	6 741[g]
25	24	25[g]
11	10	11[h]
26	27	28[h]
23	23	24[h]
101	104	108
105	103	105
...	...	...
4 052	4 456	7 960
3 247	5 684	6 135
2 792	5 563	5 944
58.7	115.4	199.7
...	116[d]	165[d]
1.053[p]	1.213[p]	1.910[p]
491	674	3 771
0.02	0.02	0.02
...	...	96
379[f]	522	553[g]
476[f]	550	645[g]
259[f]	272[c]	290[g]
...	...	...
92/92[f]	94/94[c]	94/95[h]
4.5[f]	4.9[c]	4.1[g]
750[f]	606[k]	...
3 462	3 399	3 402

Nicaragua

1976	1981	1986
2.24	2.86[a]	3.38
47.9	47.4	46.7
3.5/4.2	3.6/4.2	3.8/4.4
44/18[b]	51/13[c]	...
42.0[b]	45.4[c]	...
16.8[b]	12.1[c]	...
50.3	53.4	56.6
2.8	3.3	3.4
55/57	59/61	62/65
93	76	62
1 585[f]	...	...
5.7	...	...
658[f]	...	...
20	...	...
23	...	...
23	...	...
21	...	...
107	96	86
115	88	74
...	...	...
33	42	49
532	999	770
542	500	247
42[ef]	74[e]	...
...	...	...
7.026[q]	10.05[q]	70.00
146	111	...
0.02	0.02	...
207	...	...
17[f]	27	...
15[f]	...	16
34[f]	63[c]	58
...	...	...
54/58[f]	73/79[c]	66/86
1.9[f]	2.8[c]	5.4
1 543[f]	...	...
...	...	...

a: United Nations estimate; **b:** 1977; **c:** 1980; **d:** Domestic supply; including prod tion of exported products; **e:** Managua; metropolitan area; **f:** 1975; **g:** 1985; **h:** 19 **k:** 1982; **p:** Currency: New Zealand dollars; **q:** Currency: Cordobas.

iger			Nigeria		
1976	1981	1986	1976	1981	1986
4.73	5.46[a]	6.70[a]	70.07[a]	83.31[a]	98.52[a]
44.8	45.9	46.7	47.6	48.1	48.3
5.9/7.0	5.1/6.1	4.4/5.3	3.6/4.3	3.6/4.3	3.7/4.3
50/51[b]	...	...	47/31[b]	43/21[d]	...
...	...	...	...	...	...
...	...	...	...	...	...
10.6	13.2	16.2	18.2	20.4	23.0
2.6	2.8	3.0	3.5	3.3	3.5
39/42	41/44	43/46	45/48	47/50	49/52
157	146	135	124	114	105
842[b]	2 538[h]	...	35 413[b]	88 222[h]	...
0.0	8.8	...	6.8	1.6	...
180[b]	478[h]	...	523[b]	1 095[h]	...
18	25[h]	...	29	27	...
51	43[h]	...	22	20	...
14	17[h]	...	30	32	...
6	4[h]	...	5	5	...
78	101	108	89	103	128
86	98	91	101	100	105
...	...	...	90	84	...
...	21	35	104 515	75 881	76 018
127	510	...	8 213	20 453	6 205[c]
134	455	...	10 771	18 087	13 113[c]
62.3[e]	124.3[e]	139.1[e]	60.5[f]	120.8[f]	248.9[f]
...	...	...	...	...	...
248.5[p]	287.40[p]	322.75[p]	0.631[q]	0.637[q]	3.317[q]
83	105	189	5 180	3 895	1 081
0.00	0.01	0.01	0.57	0.69	0.69
...	21[g]	29[g]	...	...	...
2	6	7[k]	...	14	...
1	2	2[c]	...	...	3[c]
-	0.9[h]	2[c]	1[b]	5[h]	5[c]
...	...	...	...	...	...
15/8[b]	22/11[h]	23/12[c]	37/26[b]	74/50[h]	...
2.1[b]	2.8	...	...	5.3	1.1[c]
56 870[b]	37 269[n]	...	17 630[b]	9 591[h]	...
1 958	2 363	2 265	2 143	2 245	2 061

United Nations estimate; **b:** 1975; **c:** 1985; **d:** 1983; **e:** Niamey; excluding rent; **f:** ıral and urban areas; **g:** Air arrivals; **h:** 1980; **k:** 1984; **n:** 1978; **p:** Currency: CFA ıncs; **q:** Currency: Naira.

Norway

1976	1981	1986
4.03	4.10	4.17
23.8	22.2	20.1
17.1/21.1	18.0/22.5	18.7/23.5
56/37[d]	57/40	...
9.9[e]	8.1[f]	7.1[c]
24.9[e]	21.6[f]	20.2[c]
68.2	70.5	72.8
0.4	0.3	0.2
72/79	73/80	73/80
9	8	7
28 449[e]	57 713[f]	58 371[c]
4.7	4.7	3.5
7 100[e]	14 125[f]	14 092[c]
36	28	22[c]
5	4	3[c]
27	34	36[c]
20	15	14[c]
87	104	105
88	103	103
81	99	126
21 398[g]	56 395[g]	77 254[g]
11 120	15 652	20 306
7 951	18 220	18 097
72.9	113.6	165.1
...	111[h]	144[h]
5.185[s]	5.807[s]	7.400[s]
2 189	6 253	12 525
0.98	1.18	1.18
1 191[n]	1 280[n]	1 637[n]
238[e]	354	425[c]
350[e]	485	622[q]
260[ep]	292[fp]	330[cp]
...	...	...
95/95[e]	96/98[f]	96/99[q]
5.7[e]	6.0[f]	5.9[c]
580[e]	493	...
3 116	3 375	3 203

Oman

1976	1981	1986
0.80[a]	1.04[a]	2.00
44.7	44.0	44.3
4.3/4.9	3.9/4.9	3.7/4.5
...	...	...
...	...	...
...	...	...
6.1	7.3	8.8
3.2	5.0	4.7
48/51	51/54	54/57
135	117	100
2 099[e]	5 981[f]	10 019
4.0	3.1	15.1
2 740[e]	6 078[f]	5 010
36	23	28
2	2	3
60	61	52
0	1	3
...	...	...
...	...	...
...	...	...
18 290	19 525	34 514
725	2 288	2 384
1 578	4 696	2 526
...	...	...
...	...	...
0.345[kt]	0.345[kt]	0.384
307	744	968
0.05	0.27	0.29
...	...	88
19[e]	128	...
6[e]	22	44
...	...	725
...	...	...
36/14[e]	52/26[f]	74/55
2.0[e]	5.2	...
5 033[er]	2 989[fr]	1 944
...	...	...

a: United Nations estimate; **b:** Marked break in series; **c:** 1985; **d:** 1978; **e:** 1975 1980; **g:** Including Svalbard and Jan Mayen Islands; **h:** Domestic supply; **k:** Buy rate; **n:** Tourist nights in approved hotels only; **p:** Number of licences issued or s declared; **q:** 1984; **r:** In government service only; **s:** Currency: Kroner; **t:** Curren Omani rials.

akistan

1976	1981	1986
73.20[a]	85.12[a]	99.16[a]
45.5	44.4	43.6
5.0/4.6	4.7/4.5	4.5/4.5
52/4[b]	52/8[d]	52/7[e]
57.3[b]	53.9	50.7[e]
13.3[b]	14.0	14.1[e]
26.4	28.1	29.8
2.8	3.1	2.2
49/47	51/49	53/51
130	120	109
18 168[b]	28 077[c]	33 136[e]
4.5	6.6	6.4
176[b]	326[c]	330[e]
19	15	16[e]
29	26	23[e]
16	18	20[e]
14	15	16[e]
85	104	128
98	101	104
69[gh]	113[gh]	168[gh]
5 238	7 742	11 815
2 174	5 549	5 373
1 163	2 880	3 306
...	111.9	151.3
...	108[n]	140[n]
9.90[qs]	9.90[qs]	17.25[qs]
466	721	709
1.62	1.85	1.93
197	290	432
3[b]	7	6[e]
3[b]	4	6[e]
5[b]	10[c]	13[e]
...	64/85	...
38/15[b]	34/16[c]	41/19[f]
1.5[b]	1.3[c]	1.6[e]
4 023[b]	2 911[d]	...
2 108	2 221	2 186

Panama

1976	1981	1986
1.75	2.00	2.23
43.1	40.5	37.5
6.0/6.0	6.3/6.3	6.6/6.7
...	44/17[c]	...
...	26.1[c]	26.0[f]
...	11.2[c]	11.1[f]
49.1	50.5	52.4
2.3	2.2	2.1
68/71	69/73	70/74
32	26	23
1 841[b]	3 559[c]	4 881[e]
4.6	6.1	2.4
533[b]	1 819[c]	2 238[e]
31	28	15[e]
12	9	9[e]
14	14	13[e]
11	10	9[e]
86	104	110
95	102	93
79[g]	74[g]	76[g]
12	113	180
780[x]	1 392[x]	1 275[x]
228[x]	319[x]	327[x]
74.7[k]	107.3[k]	117.2[k]
...	110[p]	128[ep]
1.00[rt]	1.00[rt]	1.00[rt]
79	120	170
...	...	...
...	...	303
39[b]	73	76[e]
84[b]	95	104[e]
106[b]	112[c]	160[e]
...	14/15[c]	11/12[e]
87/87[b]	84/86[c]	82/83[e]
5.3[b]	4.7[c]	5.3[e]
1 310[b]	1 045	...
2 370	2 322	2420

Excluding Jammu and Kashmir, the final status of which has not yet been deter-ned; also excluded are Junagardh, Manavadar, Gilgit and Baltistan; **b:** 1975; **c:** 1980; 1982; **e:** 1985; **f:** 1984; **g:** Manufacturing; **h:** Fiscal year beginning 1 July; **k:** Panama ty; **n:** Domestic supply; including exported products; excluding products of mining d quarrying; **p:** Domestic supply; **q:** Fixed rate; **r:** Buying rate; **s:** Currency: rupees; Currency: Balboas; **x:** Excluding trade of the free zone of Colon.

Papua New Guinea | Paraguay

Papua New Guinea 1976	Papua New Guinea 1981	Papua New Guinea 1986	Paraguay 1976	Paraguay 1981	Paraguay 1986
2.75	3.04	3.40	2.78	3.25	3.81
42.0	43.0	41.6	44.0	42.7	41.7
5.1/5.3	3.8/3.5	4.6/4.6	4.8/5.7	5.0/5.9	5.0/5.9
58/44[a]	...	...	50/14	55/14[c]	...
...	...	...	...	42.9[c]	...
...	...	...	...	12.3[c]	...
11.9	13.0	14.3	39.0	41.7	44.4
2.7	2.6	2.4	3.3	3.0	2.8
50/50	51/53	53/55	62/66	63/68	64/69
85	74	62	49	45	42
1 400[a]	2 549[b]	2 292[k]	1 511[a]	4 448[b]	5 808
3.2	0.5	1.2	7.2	10.7	1.4
519[a]	826[b]	653[k]	562[a]	1 404[b]	1 577
18	27	21[k]	23	27	19
35	34[b]	...	35	28	29
19	23[b]	...	18	19	19
9	9[b]	...	16	17	16
90	104	113	79	104	120
103	102	98	92	98	97
...	...	...	64	104	106
20	26	37	47	60	141
430[d]	1 096[d]	...	180[d]	506[d]	733
533	863	1 038	181	296	275
76.4	108.1	144.6	52.7[f]	114.0[f]	166.1
...	...	...	...	...	...
0.81[p]	0.68[p]	0.96[p]	126.0[gq]	126.0[gq]	550.0
257	396	425	158	806	447
0.03[h]	0.06	0.06	0.00	0.04	0.04
...	35	32	...	178	371
7[a]	15	13	4[a]	19	...
13[a]	16	17[k]	10[a]	19	26
...	...	...	20[a]	21[a]	23
...	...	...	...	9/15[c]	...
45/27[a]	43/32[b]	...	66/61[a]	71/66[c]	70/66
6.9	...	...	...	...	1.3
14 495	16 052[b]	...	2 195	1 747[n]	...
...	...	...	2 725	2 780	2 813

a: 1975; **b:** 1980; **c:** 1982; **d:** F.o.b.; **e:** 1984; **f:** Asunción; **g:** Selling rate; **h:** 19
k: 1985; **n:** 1979; **p:** Currency: Kina; **q:** Currency: Guaranies.

'eru

	1976	1981	1986
	15.57	17.75	20.21
	43.2	41.8	40.5
	5.2/6.0	5.1/6.0	5.1/6.0
	45/13[a]	47/16	...
	...	34.9	...
	...	12.4	...
	61.4	64.5	67.4
	2.6	2.6	2.5
	55/59	57/61	59/63
	105	99	88
	15 377[a]	19 379[k]	14 394[d]
	4.5	1.6	-0.9
	1 014[a]	1 120[k]	731[d]
	17	19	14[d]
	13	9	8[n]
	33	35	37[n]
	27	25	25[n]
	102	101	110
	118	99	96
	99[e]	100[e]	88[de]
	5 018	11 393	10 337
	2 016	3 803	2 160
	1 341	3 249	2 467
	177[f]	175[f]	6 001[f]
	...	168[g]	5 412[g]
	...	0.507[q]	13.95[q]
	289	1 200	1 430
	1.00	1.40	2.14
	264	335	303
	17[a]	29	30[d]
	17[a]	27	32[d]
	40[a]	49[k]	76[d]
	...	10/26	...
	...	95/89	101/95[d]
	3.3[a]	2.9[k]	2.8[n]
	1 560	1 480[p]	...
	2 272	2 179	2 144

Philippines

	1976	1981	1986
	43.41	49.54	55.00
	42.8	42.0	40.4
	4.2/4.6	5.1/5.4	5.0/5.5
	46/17[a]	46/28	...
	50.2[a]	50.0	49.9[c]
	10.6[a]	11.3	10.1[c]
	35.6	37.4	39.6
	2.5	2.4	2.3
	58/61	60/64	62/65
	54	51	45
	15 825[a]	35 235[k]	32 787[d]
	6.2	6.1	-0.5
	372[a]	729[k]	602[d]
	25	26	16[d]
	28	23	26[d]
	27	28	28[d]
	24	25	25[d]
	88	104	110
	99	101	95
	80	113	231[d]
	293	939	1846
	3 953	7 946	5 394
	2 574	5 722	4 842
	61.2	113.1	255.8
	...	113[h]	291[h]
	7.428[r]	8.20[r]	20.53[r]
	1 597	2 066	1 728
	1.06	1.66	2.26
	615	923	764
	9[a]	18	17[d]
	11[a]	15	15[d]
	18[a]	21[k]	28[d]
	...	16/17[k]	...
	...	94/96[k]	90/92[d]
	1.5[a]	1.5[k]	1.2[d]
	...	...	...
	2 132	2 355	2 313

: 1975; **b:** 1978; **c:** 1983; **d:** 1985; **e:** Manufacturing; **f:** Lima; metropolitan area; **g:** omestic supply; **h:** Manila; domestic supply including exported products; excluding roducts of mining and quarrying; **k:** 1980; **n:** 1984; **p:** 1979; **q:** Currency: intis; **r:** urrency: pesos.

Poland 1976	Poland 1981	Poland 1986	Portugal 1976	Portugal 1981	Portugal 1986
34.36	35.90	37.46	9.67	9.86	10.29
24.0	24.3	25.3	27.9	25.9	24.6
11.5/15.9	10.8/15.5	11.2/16.3	12.5/16.0	13.0/16.1	13.3/17.0
58/47[a]	57/45[b]	...	59/35[c]	57/36	...
34.6[c]	...	...	27.8[c]	23.9	...
30.2[c]	...	...	25.6[c]	26.0	...
55.2	58.2	61.0	27.8	29.5	31.2
0.9	0.9	0.7	0.9	0.6	0.6
67/75	67/75	68/76	67/74	68/75	69/77
23	20	17	30	20	17
1 602[rs]	2 160[rs]	8 586[frs]	14 724[a]	25 090[k]	20 687[f]
10.0[s]	1.4[s]	0.1[s]	5.2	5.2	0.7
46 624[t]	60 167[t]	230 806[ft]	1 562[a]	2 533[k]	2 026[f]
29[q]	10[q]	18[fq]	19	31	22[f]
15	30	16[f]	13	9	...
52	42	49[f]	33	32	...
...	...	...	31	31	...
104	96	116	104	88	105
107	95	110	108	88	100
90	86	103	73	100	123
137 783	101 078	124 630	546	507	872
13 867[d]	15 224[d]	11 107[d]	4 318	9 951	9 458
11 017	13 182	11 884	1 824	4 180	7 205
75.3	121.2	463.4	44.4	120.0	317.5
...	...	...	...	122[e]	305[e]
19.92[v]	33.20[v]	151.00[fv]	31.549[x]	65.249[x]	146.117[x]
...	...	...	176	534	1 456
...	...	...	27.67	22.14	20.16
...	1 737	2 398[g]	...	3 021[h]	5 409[h]
32[a]	91	119[f]	99[a]	151	...
75[a]	97	113[f]	113[a]	149	180[f]
190[an]	224[kn]	255[fn]	77[an]	143[kn]	157[fn]
...	...	...	...	15/25	11/20[f]
89/89[a]	91/92[k]	94/94[f]	87/85[a]	78/78[k]	...
3.7[a]	4.0[k]	4.7[f]	3.8[a]	4.0	...
620[a]	541[p]	...	850[a]	456[p]	...
3 471	3 433	3 253	3 025	3 058	3 153

a: 1975; **b:** 1978; **c:** 1974; **d:** F.O.B.; **e:** Lisbon; domestic supply; including export products; excluding products of mining and quarrying; **f:** 1985; **g:** 1984; **h:** Excludi nationals residing abroad; **k:** 1980; **n:** Number of licences issued or sets declare **k:** excluding rent; **p:** 1982; **q:** Net fixed capital formation; **r:** Billion Zlotych; **s:** N material product; **t:** In national currency; **v:** Currency: Zlotych; **x:** Currency: Escudo

Qatar			Republic of Korea		
1976	1981	1986	1976	1981	1986
0.19[a]	0.25[a]	0.33[a]	35.85	38.72	41.57
33.4	32.3	34.5	37.7	34.0	31.2
2.7/3.9	2.3/2.3	2.6/2.5	5.0/6.7	5.1/6.9	5.3/7.6
66/4[b]	64/10[c]	...	47/30[d]	47/26[f]	46/28[e]
...	...	...	46.5[d]	35.3[f]	26.1[e]
...	...	...	17.5[d]	21.5[f]	23.5[e]
83.6[d]	86.0	88.0[g]	56.9	65.3	71.1
8.6	5.4	6.8	1.6	1.6	1.7
61/64	64/68	65/70	62/69	65/71	66/73
57[a]	46[a]	38[a]	35	30	24
2 469[d]	7 829[f]	6 532[g]	21 146[d]	62 419[f]	86 180[g]
5.0	1.9	-3.1	9.2	8.0	7.9
14 439[d]	34 769[f]	20 737[g]	599[d]	1 637[f]	2 089[g]
...	17	...	24	27	29[g]
1	1	...	24	16	14[g]
67	69	...	30	33	33[g]
4	5	...	27	29	28[g]
...	...	...	99	98	116
...	...	...	104	97	106
...	...	...	62	113	195
25 691	25 179	20 784	8 706	9 452	13 936
833	1 518	1 139[g]	8 774[h]	26 131[h]	31 584[h]
2 209	5 844	3 541[g]	7 715	21 254	34 714
...	100	111.9[gk]	52.1	121.3	144.2
...	...	...	...	120[n]	126[n]
3.958[q]	3.640[q]	3.640[q]	484.0[r]	700.5[r]	861.4[r]
129	366	572	1 970	...	3 320
0.19	0.71	0.97	0.11	0.30	0.32
...	109	98[g]	834	1 093	1 659
148[d]	401	...	3[d]	14	27
126[d]	276	470[g]	40[d]	90[f]	186[g]
117[d]	356[f]	381[g]	...	165[f]	187[g]
...	49/49	...	...	...	...
82/84[d]	90/89[f]	93/100	85/87[d]	95/90[f]	96/93
1.4[d]	2.5	4.3[g]	1.6[d]	3.2[f]	3.9[g]
1 167[dp]	...	...	2 100[d]	1 441	...
...	...	...	2 755	2 824	2 822

United Nations estimate; **b:** 1974; **c:** 1982; **d:** 1975; **e:** 1984; **f:** 1980; **g:** 1985; **h:** cluding imports of foreign aid; **k:** Doha; **n:** Domestic supply; **p:** Physicians in government services only; **q:** Currency: Qatar riyals; **r:** Currency: Won.

Romania 1976	Romania 1981	Romania 1986	Rwanda 1976	Rwanda 1981	Rwanda 1986
21.45	22.35	23.17[a]	4.29	5.35	6.27[a]
25.2	26.7	25.2	47.7	47.7	48.1
12.7/15.8	11.6/14.9	12.4/16.0	3.9/4.7	3.7/4.5	3.7/4.5
55/45[b]	...	...	56/51[c]	...	...
...	...	...	...	...	...
...	...	...	...	...	...
46.2	48.1	49.0	4.0[c]	5.0	6.2[h]
0.9	0.7	0.7	3.2	3.3	3.3
67/72	68/73	68/73	43/46	43/47	45/48
31	26	23	140[a]	140[a]	132[a]
...	...	...	568[c]	1 163[g]	1 587[n]
...	...	...	9.0	5.3	2.2
...	...	...	130[c]	226[g]	270[n]
...	...	...	13	13	14[n]
...	...	...	49	41	43[n]
...	...	...	15	16	17[n]
...	...	...	12	16	16[n]
97	98	125	84	107	107
100	98	121	97	102	87
71	103	122	...	...	...
57 481	61 399	63 010	13[c]	9	15
6 095[d]	10 978[d]	10 590[d]	103	256	352
6 138	11 180	12 543	81	88	118
93.9	102.0	...	...	...	...
...	...	...	...	...	...
20.00[p]	15.00[p]	15.28[p]	92.84[q]	92.84[q]	84.18[q]
562	404	582	64	173	162
2.75	3.59	3.25	...	...	...
3 169[e]	7 002[e]	4 535[e]	...	...	...
...	17	...	1[c]	...	3[h]
56[c]	...	...	1[c]	1	2[h]
127[cf]	167[fg]	173[fh]	...	...	...
...	...	...	...	49/73[k]	...
93/93[c]	94/92[g]	90/90[h]	35/29[c]	43/39[g]	43/40[h]
3.0[c]	2.9[g]	2.1[n]	2.3[c]	3.8	3.0[n]
760	678[g]	...	41 900	30 768	28 071[n]
3 310	3 352	3 394	1 923	2 073	2 013

a: United Nations estimate; **b:** 1977; **c:** 1975; **d:** F.O.B.; **e:** Visitor arrivals; **f:** Numb of licences issued or sets declared; **g:** 1980; **h:** 1985; **k:** 1978; **n:** 1984; **p:** Curren Lei; **q:** Currency: Rwanda francs.

aint Kitts nd Nevis			Saint Lucia		
1976	1981	1986	1976	1981	1986
0.05	0.04	0.05[a]	0.11	0.13	0.13[a]
...	...	...	...	...	...
...	...	...	...	...	...
...	...	...	...	...	...
...	...	...	...	...	...
...	...	...	...	...	...
37.7[b]	41.3	43.5[h]	...	...	...
-0.1	-1.7	2.3	2.1	1.6	1.2
...	...	...	...	...	...
...	46	...	30	24	18[p]
34[b]	48[g]	65[h]	57[b]	113[g]	151[p]
3.7	2.3	4.3	1.2	8.9	3.9
667[b]	923[g]	1 226[h]	514[b]	942[g]	1 277[p]
...	31	32[p]	41	40	26[p]
17	9	12[p]	13	9	12[p]
14	13	14[p]	10	11	13[p]
13	12	14[p]	7	8	9[p]
...	...	...	...	...	...
...	...	...	...	...	...
...	...	...	...	...	...
...	...	...	...	...	...
24[b]	48	...	46[b]	118[c]	...
22[b]	24	...	16[b]	42[c]	...
...	110.4[d]	125.7[d]	63.4	115.1	102.2[e]
...	...	...	...	...	...
2.372[bfq]	2.70[fq]	2.70[fq]	2.372[q]	2.70[q]	2.70[q]
...	...	...	...	8	25
...	...	...	...	...	...
...	35	57	...	69	112
...	80	90[p]	...	52	64[p]
36[b]	...	...	54[b]	69[g]	86[p]
...	63[g]	109[h]	15[b]	15[g]	17[h]
...	...	...	...	...	...
...	...	...	...	...	...
4.0[b]	5.0[g]	6.9[h]	6.8[b]	6.3[g]	...
3 333[n]	2 609	...	2 900[n]	3 215	...
2 240	2 264	2 233	2 128	2 312	2 421

United Nations estimate; **b:** 1975; **c:** 1982; **d:** St. Kitts; **e:** Base: 1985 = 100; **f:** Operanal exchange rate for United Nations programmes as of 1 October; **g:** 1980; **h:** 1985; 1977; **p:** 1984; **q:** Currency: East Caribbean dollars.

St. Vincent and the Grenadines | Samoa

St. Vincent and the Grenadines 1976	1981	1986	Samoa 1976	1981	1986
0.09[a]	0.10	0.10[a]	0.15	0.16	0.16[a]
...	...	...	...	...	...
...	...	...	...	...	...
...	...	...	41/9	44/8	...
...	...	...	61.1	60.4	...
...	...	...	3.1	2.9	...
...	...	...	...	...	...
0.7	1.3	-4.0	...	...	...
...	...	...	...	...	...
...	60[e]	37[f]	...	...	...
...	58[e]	102[d]	...	112[e]	88
...	6.2	5.5	...	3.1	-0.6
...	586[e]	981[d]	...	723[e]	540
36[h]	33	31[f]	...	...	...
16[h]	14	14[f]	...	...	...
10[h]	12	11[f]	...	...	...
6[h]	9	8[f]	...	...	...
...	...	...	...	...	...
...	...	...	...	...	...
...	...	...	...	...	...
...	2[b]	2	...	...	...
25[c]	58	...	30	56	48
8[c]	24	...	7	11	12
...	...	...	...	120.5[eq]	202.5
...	...	...	...	...	...
2.372[k]	2.70[k]	2.70[k]	0.80[p]	1.099[p]	2.198
...	9	26	5	3	24
...	...	...	...	...	...
...	40	42	...	...	47
...	64[e]	...	9[c]	11	24
48[c]	50	65[f]	...	...	...
...	...	58[d]	-	16[e]	31
...	...	...	...	...	...
...	...	...	...	...	...
...	...	...	...	...	...
...	7 146[e]	...	2 745[c]	2 492	...
2 188	2 459	2 684	2 217	2 403	2 373

a: United Nations estimate; **b:** 1983; **c:** 1975; **d:** 1985; **e:** 1980; **f:** 1984; **h:** 1977;
Currency: East Caribbean dollars; **p:** Currency: Tala; **q:** Excluding rent.

ao Tome nd Principe | Saudi Arabia

1976	1981	1986	1976	1981	1986
0.08[a]	0.09[a]	0.11[a]	7.62[a]	9.80[a]	12.01[a]
...	...	...	44.3	44.2	44.6
...	...	...	4.4/5.3	3.9/4.9	3.8/4.8
...	43/20	...	50/3[b]	...	...
...	53.9	...	...	...	...
...	6.2	...	...	...	...
27.9[b]	32.8	...	58.7	65.9	72.4
2.7	1.2	...	5.1	4.2	3.8
...	...	...	56/60	59/63	62/66
60	70[g]	62	100	85	71
...	...	...	46 767[b]	156 474[f]	79 152[d]
...	...	...	11.6	8.6	-4.8
...	...	...	6 450[b]	16 696[f]	6 858[d]
...	...	...	25	23	...
...	...	...	1	1	...
...	...	...	68	66	...
...	...	...	5	4	...
...	...	...	103	87	259
...	...	...	126	82	202
...	...	...	...	...	...
0	1	0	432 792[c]	505 651[c]	268 888[c]
11[b]	18	...	8 694	35 244	23 623[d]
7[b]	25	...	38 287	120 240	27 480[d]
...	...	...	76.6[e]	102.5[e]	96.0[e]
...	...	...	...	...	...
31.549[p]	38.876[p]	36.993[p]	3.530[q]	3.415[q]	3.745[q]
...	...	...	26 900	32 236	18 324
...	...	...	3.08	4.57	4.60
...	...	...	...	...	...
...	23	23[d]	10[b]	192	...
...	...	...	...	112	118[d]
...	...	...	...	224[f]	269[d]
...	27/58	...	...	29/69[g]	...
...	...	...	52/30[b]	58/38[f]	66/49[d]
...	...	...	4.4[b]	3.0[f]	6.2[d]
1 952[h]	2 263[f]	...	2 010[k]	1 673	...
1 953	2 353	2 435	2 004	2 826	3 093

United Nations estimate; **b:** 1975; **c:** Part of the neutral zone; **d:** 1985; **e:** All cities; 1980; **g:** 1982; **h:** 1977; **k:** 1978; **p:** Currency: Dobras; **q:** Currency: Saudi Arabian als.

Senegal | Seychelles

Senegal 1976	Senegal 1981	Senegal 1986	Seychelles 1976	Seychelles 1981	Seychelles 1986
5.11	5.87	6.61[a]	0.06	0.06	0.07
43.8	44.1	44.4	...	...	...
4.5/5.2	4.5/5.1	4.5/5.1	...	...	...
53/33	...	...	53/31[b]	49/30	49/36
...	...	...	19.5	9.8	...
...	...	...	6.4	8.4	...
34.2	34.9	36.4	33.2	42.8	...
3.5	2.6	2.7	3.4	1.3	0.6
40/43	42/45	44/47	...	...	...
154	142	131	...	19[e]	14
1 896[c]	2 970[g]	2 642[d]	...	147[g]	152
1.9	0.6	4.0	...	8.2	-0.3
397[c]	524[g]	410[d]	...	2 262[g]	2 079
13	15	85[d]	37	34	21
30	18	18[d]	10	8	7
18	19	21[d]	7	10	11
...	...	...	5	9	9
122	122	118	...	...	...
140	118	100	...	...	...
110	109	134	...	...	...
...	...	...	...	...	...
640	861	620[d]	...	...	...
476	500	402[d]	...	...	...
72.9[f]	105.9[f]	186.1[f]	70.0	110.6	122.3
...	...	...	...	...	...
248.5[p]	287.40[p]	322.75[p]	7.832[q]	6.227[q]	5.929
25	9	9	6	14	8
0.00	0.03	0.03	...	...	...
136	216	235	49	60	67
10[c]	9[g]	...	42[c]	...	69
8[c]	...	...	52[c]	113	173
0.4[c]	0.7[g]	31[d]	-	-	26
...	...	...	...	...	...
...	36/22[g]	43/27[d]	...	...	...
4.0[c]	4.7[g]	...	3.9[c]	8.1[e]	...
16 210	13 069	...	...	2 333[h]	...
2 291	2 389	2 339	2 143	2 299	2 289

a: United Nations estimate; **b:** 1977; **c:** 1975; **d:** 1985; **e:** 1982; **f:** Dakar; **g:** 19
h: 1979; **k:** 1984; **p:** Currency: CFA francs; **q:** Currency: Seychelles rupees.

ierra Leone			Singapore		
1976	1981	1986	1976	1981	1986
3.09[a]	3.35[a]	3.67[a]	2.29	2.44	2.59
40.7	40.9	41.4	32.8	27.1	24.5
4.6/5.5	4.6/5.5	4.6/5.5	6.3/7.2	6.6/7.8	7.1/8.4
51/27	...	...	53/24[b]	59/33[c]	60/34[d]
...	...	...	2.0[b]	1.5[c]	0.7[d]
...	...	...	26.3[b]	29.9[c]	25.2[d]
21.1	24.5	28.3	100	100	100
1.6	1.8	1.9	1.3	1.2	1.1
31/33	32/35	34/37	69/73	69/75	70/76
191	180	169	13	10	9
682[b]	1 231[c]	...	5 670[b]	11 719[c]	17 510[d]
1.8	3.7	...	9.8	8.7	6.5
224[b]	373[c]	...	2 506[b]	4 853[c]	6 843[d]
11	13	...	36	44	42[d]
38	34	...	2	1	1[d]
17	15	...	27	31	27[d]
7	9	...	25	29	25[d]
97	101	111	97	108	96
107	100	101	102	107	90
...	...	...	63[e]	110[e]	116[e]
...	...	...	...	...	...
153	312	132	9 070	27 608	25 512
101	153	145	6 585	20 967	22 495
64.5[f]	123[f]	1 403[f]	82.0	108.3	115.7
...	...	...	...	104[g]	90[dg]
1.175[p]	1.174[p]	35.587[p]	2.455[q]	2.048[q]	2.175[q]
25	16	14	3 364	7 549	12 939
...	...	...	...	...	...
26	53	194	1 162	2 829	2 902
5[b]	10	9[k]	66[b]	109	138[d]
4[b]	...	...	141[b]	317	417[d]
2.6[b]	6[c]	8.3[d]	124[bh]	165[ch]	195[dh]
...	...	...	...	8/26[c]	...
33/20	48.30	...	80/77[b]	83/83	93/92
3.1[b]	3.6[c]	...	2.5[b]	2.6	...
...	18 284[c]	...	1 395[b]	1 111[n]	...
1 943	2 049	1 834	2 619	2 667	2 729

United Nations estimate; **b:** 1975; **c:** 1980; **d:** 1985; **e:** Manufacturing; **f:** Freetown; Domestic supply; excluding products of mining and quarrying; **h:** Number of licences sued or sets declared; **k:** 1984; **n:** 1982; **p:** Currency: Leones; **q:** Currency: Singapore llars.

Solomon Islands 1976	Solomon Islands 1981	Solomon Islands 1986	Somalia 1976	Somalia 1981	Somalia 1986
0.20	0.23[a]	0.28[a]	3.36[a]	4.16[a]	4.76
...	...	...	42.8	44.1	44.8
...	...	...	4.8/5.6	4.7/5.5	4.6/5.4
...	...	...	56/23	...	...
...	...	...	...	...	...
...	...	...	...	...	...
9.1[g]	9.3	9.6[b]	26.5	30.2	34.1
3.5	4.0	3.9	4.2	2.9	2.1
...	...	...	39/42	39/42	40/43
...	...	...	155	155	149
67[g]	144[k]	...	697[g]	1 586[k]	...
4.9	10.8	...	2.2	1.1	...
345[g]	640[k]	...	214[g]	395[k]	...
...	...	...	16	8	...
...	...	...	51	43	...
...	...	...	8	7	...
...	...	...	7	5	...
70	109	129	94	102	108
78	106	103	111	99	91
...	...	...	...	...	...
...	...	...	...	...	...
26	76	63	156	512	112
24	66	67	95	152	91
92.2[f]	116.4[f]	113.6[ef]	...	144.4[d]	464.1
...	...	...	...	...	...
0.920[p]	0.889[p]	1.986[p]	6.295[q]	6.295[q]	90.5
...	22	30	85	31	13
...	...	...	0.00	0.02	0.02
...	...	...	...	...	...
5	...	...	...	...	...
...	...	...	...	...	...
...	...	...	-	-	0.2
39/52[g]	...	...	...	...	...
...	...	...	33/17[g]	33/17	...
...	2.4[h]	2.2[c]	1.8[g]	1.3	...
5 135	6 553[h]	...	...	15 589	...
2 046	2 119	2 085	1 975	2 054	2 059

a: United Nations estimate; **b:** 1985; **c:** 1984; **d:** Mogadishu; **e:** Base: 1985 = 100 Honiara; **g:** 1975; **h:** 1982; **k:** 1980; **p:** Currency: Solomon Islands dollars; **q:** Curr[illegible]cy: Somali Shillings.

outh Africa 1976	1981	1986	Spain 1976	1981	1986
26.13	29.31a	33.22a	35.97	37.76	38.67
41.6	41.3	41.0	27.6	26.6	24.3
6.0/6.9	5.9/6.7	5.8/6.6	12.5/16.2	12.8/16.9	13.6/17.8
50/25c	46/23b	...	52/21c	52/17	...
...	15.0b	...	22.3c	13.2	15.0d
...	27.4b	...	26.6c	23.1	22.9d
50.5	53.2	55.9	69.6	72.8	75.8
2.3	2.5	2.5	1.0	0.6	0.6
50/53	52/55	54/57	71/76	71/77	71/78
95	83	72	16	10	9
37 505c	79 701b	54 834d	104 836c	212 115b	164 254d
4.1	2.5	0.9	6.0	1.9	1.5
1 471c	2 786b	1 693d	2 945c	5 667b	4 262d
30	28	24d	22	22	19d
7	7	5d	9	6	6d
36	40	39d	29	31	30d
22	22	20d	25	...	...
85	109	99	90	94	105
93	108	85	94	93	101
81e	106	95e	91	99	107
42 681f	72 277f	93 610f	12 341	24 742	21 582
6 738fg	21 077fg	11 980fg	17 463	32 218	35 022
5 212h	11 076h	10 861h	8 727	20 351	27 158
62.9k	115.2k	220.5k	50.2	114.6	193.7
...	...	...	...	116n	182n
0.870s	0.957s	2.183s	68.288t	97.450t	132.395t
425	666	370	4 704	10 805	14 755
12.67	9.29	4.82	14.27	14.61	14.82
...	709	645	30 014	24 742	29 910
83c	119	132d	135c	248	282d
78c	118	143d	220c	329	363d
...	70b	93d	187c	252b	270d
...	...	...	4/10c	4/10	...
...	...	...	90/89c	96/97b	96dq
...	...	...	...	...	...
1 906r	...	...	650	390	...
2 907	2 932	2 945	3 624	3 337	3 335

United Nations estimate; **b:** 1980; **c:** 1975; **d:** 1985; **e:** Manufacturing; **f:** Customs ea comprising South Africa, Botswana, Lesotho, Namibia and Swaziland; **g:** F.O.B.; Excluding exports of gold; **k:** White population; **n:** Domestic supply; excluding pro- cts of electricity, gas and water; **q:** Both sexes; **r:** 1978; **s:** Currency: Rand; **t:** Cur- ncy: Pesetas.

Sri Lanka			Sudan		
1976	1981	1986	1976	1981	1986
13.72	15.01	16.12	16.13[a]	19.24[a]	22.18
39.4	35.3	34.5	44.4	44.9	45.1
6.6/5.6	6.9/6.4	7.3/6.8	4.2/4.8	4.2/4.8	4.2/4.9
50/17[b]	65/23	...	...	...	...
...	37.4	...	...	...	...
...	9.1	...	...	...	...
22.0	21.6	21.1	18.9	19.7	20.6
1.7	1.8	1.5	3.1	2.9	2.9
65/68	67/70	68/72	44/46	47/49	49/52
48	39	33	131	118	106
3 859[b]	4 133[g]	5 808[d]	5 310[b]	9 901[g]	...
3.3	5.6	5.1	1.8	2.6	...
284[b]	279[g]	358[d]	332[b]	530[g]	...
14	28	24[d]	13	12	...
28	27	24[d]	35	34	...
25	19	19[d]	10	8	...
24	17	17[d]	8	7	...
81	97	103	86	111	123
81	94	91	97	110	104
76	107	...	...	...	...
96	133	227	34	42	44
547	1 849	1 793	980	1 578	757
567	1 044	1 099	554	658	367
63.1	118.0	190.4	43.5	124.3	203.9
...	123[f]	159[f]	...	...	...
8.828[p]	20.550[p]	28.520[p]	0.348[q]	0.901[q]	2.50
92	327	353	24	17	59
...	0.06	0.06	...	...	...
119	371	257	25	22	42
7[b]	16	18[d]	...	4	4
5[b]	7	...	3[b]	4	4
-	2.4[g]	28[d]	6.2[b]	43[g]	51
...	9/18	...	...	...	...
62/60[b]	77/77[g]	82/84[d]	41/22[b]	42/29	42/30
2.6[b]	2.6[g]	2.4[d]	...	4.4[g]	4.0
6 387[b]	...	...	8 721[k]	...	...
2 061	2 228	2 410	2 142	2 319	2 003

a: United Nations estimate; **b:** 1975; **c:** Manufacturing; **d:** 1985; **e:** 1983; **f:** Domes supply; **g:** 1980; **h:** 1984; **k:** 1977; **p:** Currency: Sri Lanka rupees; **q:** Curren Sudanese pounds.

uriname 1976	1981	1986	Swaziland 1976	1981	1986
0.35	0.36	0.38a	0.49	0.57	0.67
47.6	39.8	37.2	44.9	45.4	46.0
5.7/5.9	6.1/6.5	6.3/6.8	4.4/5.2	4.4/5.2	4.4/5.2
37/12b	...	...	52/42b	...	...
...	7.8	...	...	...	...
...	14.4	...	...	...	...
44.8	44.8	45.7	14.0	19.8	26.3
-0.5	1.1	1.5	2.9	3.0	3.1
64/69	66/71	67/72	45/48	47/50	49/52
44	36	30	140	129	118
521b	898f	963k	291b	598f	...
9.1	2.3	-1.4	15.3	6.3	...
1 427b	2 530f	2 603k	602b	1 070f	...
31	27	18k	22	25	...
7	8	7k	...	20	...
29	26	22k	...	20	...
18	15	12k	...	16	...
87	103	129	83	107	113
85	103	122	95	104	94
...	...	...	...	...	...
99	71	212	c	c	c
294	568	299g	c	c	c
276	474	...	c	c	c
66.0e	108.7e	166.2e	57.4d	119.8d	224.7d
...	...	...	...	...	...
1.785q	1.785q	1.785q	0.870r	0.957r	2.183r
110	207	21	73	96	96
0.15	0.05	0.05	...	...	...
55	40	20	...	83	256
70b	84	...	15b	37	...
44b	...	100g	15b	23	27g
93b	113f	117g	-	1.8f	12g
32/37np	...	...	43/46	...	...
84/84b	76fh	87gh	77/72b	85/84	87/86k
5.5b	6.7f	...	...	4.0	...
1 892b	1 728n	...	7 950b	7 200n	...
2 291	2 553	2 666	2 456	2 497	2 562

United Nations estimate; **b:** 1975; **c:** Included with data for South Africa; **d:** Mbabane-anzini, excluding rent; **e:** Paramaribo; **f:** 1980; **g:** 1985; **h:** Both sexes; **k:** 1984; **n:** 78; **p:** Age group: 15-59; **q:** Currency: Guilders; **r:** Currency: Emalangeni.

Sweden 1976	Sweden 1981	Sweden 1986	Switzerland 1976	Switzerland 1981	Switzerland 1986
8.22	8.32	8.37	6.35	6.35	6.50
20.7	19.6	18.2	22.3	19.7	17.6
19.2/22.8	19.8/24.0	20.5/25.0	15.1/19.8	15.7/20.7	16.6/21.7
55/39[a]	54/43[b]	...	64/33[a]	63/35[b]	...
6.3[a]	5.6[b]	4.7[c]	...	7.2[b]	...
28.5[a]	25.2[b]	23.1[c]	...	33.3[b]	...
82.7	83.1	83.4	55.7	57.0	58.2
0.3	0.1	-0.1	-0.2	0.1	0.0
72/78	73/79	74/80	72/79	73/80	73/80
8	7	6	10	8	7
72 443[a]	124 137[b]	100 247[c]	54 303[a]	101 629[b]	92 776
2.8	1.3	2.0	1.3	1.7	1.1
8 842[a]	14 936[b]	12 003[c]	8 478[a]	16 063[b]	14 555
21	19	19[c]	21	24	24
4	3	3[c]	2[a]	6	7
27	23	24[c]	21[a]	38	37
25	20	21[f]	36[a]	36	34
95	103	103	95	99	108
96	103	103	96	98	105
102[d]	98[d]	110[d]	88	99	108
6 011	8 266	11 271	2 892[e]	4 194[e]	4 646
19 333	28 845	32 514	14 774	30 696	41 278
18 435	28 664	37 221	14 845	27 042	37 674
66.9	112.1	160.3	90.7	106.5	124.3
...	111[f]	154[f]	...	106[g]	111
4.126[hs]	5.571[hs]	6.819[hs]	2.450[t]	1.798[t]	1.623
2 255	3 601	6 551	9 606	13 979	21 786
5.79	6.07	6.07	83.28	83.28	83.28
...	673	824	7 609	9 383	11 400
337[a]	372	405[c]	280[a]	406	437
661[a]	828	...	613[a]	751	832
355[an]	381[bn]	390[cn]	320[a]	364[b]	400
...	...	...	...	...	...
88/92[a]	90/95[b]	90[cp]	...	...	...
6.4[a]	7.8[b]	6.9[c]	4.1[a]	4.5[b]	4.4
580	478[q]	...	560[a]	410	...
3 149	3 068	3 053	3 331	3 494	3 440

a: 1975; **b:** 1980; **c:** 1985; **d:** Excludes electricity, gas and water; **e:** Includes d for Liechtenstein; **f:** Domestic supply; including exported products; excluding produ of electricity, gas and water; **g:** Domestic supply; excluding products of mining a quarrying; **h:** Selling rate; **k:** Buying rate at Zurich; **n:** Number of licences issued sets declared; **p:** Both sexes; **q:** 1982; **r:** 1984; **s:** Currency: Krona; **t:** Currency: Sw francs.

yrian Arab Republic			Thailand		
1976	1981	1986	1976	1981	1986
7.63	9.00	10.61	42.96	47.49	52.09
48.5	47.5	48.3	43.1	39.8	36.2
5.1/5.5	4.5/4.9	4.1/4.5	4.4/5.2	4.9/6.0	5.1/6.2
39/11[a]	41/8[b]	39/6[c]	42/26[a]	55/51[e]	...
48.8[a]	31.9[b]	29.9[c]	61.9	66.0[e]	...
12.3[a]	17.0[b]	14.3[c]	11.6[d]	8.3[de]	...
45.4	47.4	49.5	15.2	17.3	19.8
3.4	3.5	3.7	2.3	2.0	1.6
58/62	61/64	63/67	59/63	61/65	62/66
70	59	48	56	48	39
5 598[a]	13 062[n]	20 267[p]	14 663[a]	33 450[n]	38 343[p]
13.1	4.7	1.7	6.4	7.7	5.1
753[a]	1 484[n]	1 929[p]	354[a]	719[n]	746[p]
31	23	24[p]	22	24	22[p]
19	19	22[p]	31	24	17[p]
20	20	14[p]	21	23	25[p]
8	8	6[p]	19	20	20[p]
80	110	124	88	105	120
89	108	99	98	104	105
84	116	147	...	...	...
10 251	8 789	9 652	470	1 031	7 258
1 979	5 040	2 728	3 572	9 951	9 138
1 065	2 103	1 325	2 980	7 038	8 753
68.4[f]	118.2[f]	250.5[f]	64.1[g]	113.4[g]	130.7[g]
...	...	...	...	110[h]	109[h]
3.925[r]	3.925[r]	3.925[r]	20.40[s]	23.00[s]	26.13[s]
293	291	...	1 798	1 732	2 804
0.79	0.83	0.83	2.34	2.49	2.49
...	347	618	1 098[k]	1 979[k]	2 818[k]
7[a]	15	22[p]	6[a]	18	21[q]
22[a]	46	59[p]	7[a]	11	15[p]
30[a]	43[n]	57[p]	...	...	97[p]
...	...	...	...	8/16[n]	...
86/55[a]	88/65[n]	97/78[p]	65/59[a]	67/64[n]	...
...	...	...	2.6[a]	2.8	...
3 100[a]	2 236	...	8 365[a]	6 852	...
2 498	2 880	3 198	2 268	2 405	2 440

1975; **b:** 1979; **c:** 1983; **d:** Including sanitary services; **e:** 1982; **f:** Damascus; **g:** angkok; metropolitan area; **h:** Domestic supply; agricultural and manufacturing products; including exported products; **k:** Including nationals residing abroad; **n:** 1980; 1985; **q:** 1984; **r:** Currency: Syrian pounds; **s:** Currency: Baht.

Togo 1976	Togo 1981	Togo 1986	Tonga 1976	Tonga 1981	Tonga 1986
2.29	2.63[a]	3.05[a]	0.09	0.09	0.11[a]
44.0	44.4	44.8	44.4[b]	...	...
4.6/5.5	4.7/5.5	4.6/5.5	5.0/5.2[b]	...	...
...	48/35[c]	...	20/4	...	...
...	...	...	44.5	...	...
...	...	...	2.4	...	...
15.8	18.8	22.1	28.0[g]	31.8	34.3[d]
2.5	2.9	3.1	-0.1	2.0	2.4
46/50	49/52	51/54	...	...	...
111	102	93	...	...	...
599[g]	1 131[c]	...	33[g]	60[c]	61[d]
1.6	4.4	...	2.6	4.3	7.4
266[g]	443[c]	...	375[g]	619[c]	560[d]
25	20	...	20	24	...
27	27	...	40	33	...
17	17	...	8	9	...
7	6	...	4	7	...
86	101	103	102	105	95
96	98	84	110	102	83
...	...	...	...	...	...
0	2	0	...	...	...
186	433	288[d]	14	40	38
104	211	190[d]	4	7	5
68.0[e]	120.6[e]	137.8[e]	60.1[f]	114.9[f]	203.9[f]
...	...	...	...	...	...
248.50[p]	287.40[p]	222.75[p]	0.700[hq]	0.865[hq]	1.58[h]
67	152	333	...	...	...
0.00	0.01	0.01	...	...	...
...	117	99	...	13	16
...	13	...	5[g]	...	26[d]
...	...	...	...	...	40[k]
...	3.9[c]	5.1[d]	...	...	...
...	53/81	...	3/5	...	...
81/40[g]	90/52[n]	77/43[d]	...	...	...
3.4[g]	5.4[n]	5.1[d]	2.9[g]	3.6	4.5[d]
28 680[g]	19 477[c]	...	3 259[g]	2 853[c]	2 463[k]
2 090	2 217	2 202	3 065	2 851	2 870

a: United Nations estimate; **b:** Official estimates; **c:** 1980; **d:** 1985; **e:** Lomé; **f:** E
cluding rent; **g:** 1975; **h:** Operational exchange rate for United Nations programm
as of 1 October; **k:** 1984; **n:** 1982; **p:** Currency: CFA francs; **q:** Currency: Pa'ang

Trinidad and Tobago 1976	1981	1986	Tunisia 1976	1981	1986
1.02	1.13	1.20a	5.77	6.57	7.23ab
37.4	34.4	32.9	43.8	42.0	39.2
6.1/7.4	7.5/8.5	7.4/8.4	6.4/5.2	7.4/6.0	7.1/6.2
50/23c	52/21e	...	49/13g	45/12e	...
13.3	9.8e	9.5f	32.4g	30.5e	...
19.2d	18.8e	23.5df	17.3g	27.9be	...
48.4	56.9	63.9	47.6	52.3	56.8
1.6	1.6	1.6	2.6	2.0	2.2
65/70	66/71	68/73	58/59	60/61	63/64
26	24	20	102	85	71
2 442g	6 233e	7 558f	4 332g	8 742e	8 214f
3.0	7.9	-4.2	8.0	6.1	3.9
2 420g	5 672e	6 378f	772g	1 368e	1 160f
23	26	21f	29	31	25f
4	2	3f	18	14	15f
48	42	31f	19	26	24f
14	7	9f	10	12	12f
135	95	103	100	102	120
145	93	94	111	99	106
83h	89h	121h	69	104	...
12 546	12 713	11 493	3 950	5 836	5 664
2 010	3 109	1 330	1 529	3 791	2 901
2 214	3 764	1 376	788	2 503	1 760
60.2	114.3	195.4	...	109.0	123.8
...	...	...		113k	170k
2.40q	2.40q	3.60q	0.431r	0.516r	0.840r
1 014	3 348	474	366	536	305
0.03n	0.05	0.05	0.13	0.19	0.19
159	187	191	978	2 151	1 502
94f	165	...	21f	40	...
63f	75	...	23f	40	37f
104g	192e	270f	34g	47e	56f
...	3/7e	...	49/75g	39/68e	...
81/85g	86/90p	85/89f	72/46g	74/53e	85/69f
2.7g	2.86	5.2f	4.6g	4.7e	4.4f
1 970	1 488e	...	4 630g	3 647	...
2 650	2 853	2 967	2 613	2 772	2 827

United Nations estimate; **b:** Marked break in series; **c:** 1974; **d:** Excluding electric, gas and water; **e:** 1980; **f:** 1985; **g:** 1975; **h:** Manufacturing; **k:** Domestic supply; including exported products; excluding products of mining and quarrying; **n:** 1977; **p:** 1982; **q:** Currency: Trinidad and Tobago dollars; **r:** Currency: Tunisian dinars.

Turkey

1976	1981	1986
40.91	45.37	50.30
40.1	38.6	36.4
6.6/7.2	5.7/7.2	5.9/6.9
51/30[e]	54/31[f]	...
64.2[e]	57.8[f]	...
8.4[e]	11.1[f]	...
41.6	43.8	45.9
2.1	2.1	2.1
58/63	60/63	63/66
120	92	76
35 949[e]	5 698[f]	52 701[g]
7.1	2.6	4.8
898[e]	1 280[f]	1 069[g]
23	19	...
27	21	18[g]
21	28	32[g]
18	24	25[g]
93	102	115
100	100	103
87	109	139[g]
9 106	10 833	18 636
5 129	8 932	11 191
1 960	4 703	7 398
...	131.4[k]	380.5[k]
	137[np]	462[n]
17.0[x]	133.623[x]	757.790[x]
978	928	1 465
3.57	3.77	3.84
1 336[s]	1 052[s]	2 079[s]
10[e]	25	28[g]
25[e]	47	67[g]
...	79[f]	...
23/57[e]	19/50[f]	14/37[g]
...	79/65[t]	83/68[g]
...	2.3[f]	2.0[g]
1 850[e]	1 527[p]	...
2 965	3 104	3 180

Tuvalu

1976	1981	1986
6[ab]	7[ab]	
...	31.8[cd]	...
...	6.8/9.2[cd]	...
...	...	...
...	...	...
...	...	...
...	...	...
0.7	3.1	3.4
...	...	...
...	...	...
...	...	...
...	...	...
...	...	...
...	...	...
...	...	...
...	...	...
...	...	...
...	...	...
...	...	...
...	...	...
...	...	...
...	...	...
...	...	...
...	111.1[h]	142.2
...	...	...
0.795[eqz]	0.847[fqz]	1.208
...	...	...
...	...	...
...	...	...
...	...	...
...	...	...
...	...	...
...	...	...
...	...	...
...	...	...
2 367[v]	1 875[f]	...
...	...	...

a: United Nations estimate; **b:** In thousands; **c:** Official estimates; **d:** 1979; **e:** 197
f: 1980; **g:** 1985; **h:** Funafuti; **k:** Index base: 1982 = 100; **n:** Base: 1981 = 100; Domes
supply; including exported goods, excluding products of electricity, gas and water a
industrial finished goods; **p:** 1982; **q:** Operational exchange rate for United Natio
programmes; **r:** 1984; **s:** Visitor arrivals; **t:** 1983; **v:** 1977; **x:** Currency: Liras; **z:** C
rency: Australian dollars.

Uganda 1976	Uganda 1981	Uganda 1986	USSR 1976	USSR 1981	USSR 1986
11.94	13.54[a]	16.02[a]	256.76	267.72	280.14
47.4	47.8	48.1	26.1	24.3	24.8
3.9/4.4	3.8/4.4	3.8/4.4	9.1/17.0	8.7/17.0	8.7/17.1
56/28	...	...	44.6	47.6	47.0
...	...	...	23[f]	20	19
...	...	...	29.2	29.0	29.2
8.2	8.7	9.5	60.7	63.4	65.6
3.2	3.3	3.5	0.9[fr]	0.8[r]	0.9[r]
46/50	47/51	49/53	65/74	65/74	64/73[r]
114	112	103	30[r]	27[r]	25[r]
...	...	...	386[np]	487[np]	587[np]
...	...	...	...	4.8[p]	3.8[p]
...	...	...	1 509[q]	1 819[q]	2 107[q]
...	...	...	27	24	26
...	...	...	17	15	21
...	...	...	52	51	44
...	...	...	...	...	...
111	109	153	97	99	117
126	106	125	101	98	107
...	...	...	86	103	126
59	46	56	1 151 953	1 356 228	1 581 100
158	...	...	38 106[c]	72 960[c]	88 871[c]
359	242	399[b]	37 176	79 003	97 336
...	...	...	00.0	101.3	106.5
...	...	...	...	...	...
8.314[s]	85.150[s]	1 400.0[s]	0.750[t]	0.720[t]	0.684[t]
45	17[d]	29	...	...	...
...	...	...	...	...	...
...	4	35	3 879	5 870	2 036
2.4[f]	3.2	...	16[fh]	30	47
4[f]	3	...	69	94	110
6[f]	6[d]	6[g]	223	250[d]	299
...	...	...	...	...	...
34/22[f]	39/28	...	95[fk]	101[k]	103[gk]
2.4	1.3	...	8.4[r]	8.0[r]	7.0[r]
26 060[f]	22 322	...	296	258	232
2 324	2 169	2 291	3 388	3 385	3 403

United Nations estimate; **b:** 1984; **c:** F.O.B.; **d:** 1980; **e:** Net fixed capital formation; **f:** 1975; **g:** 1985; **h:** Passenger cars only; **k:** Both sexes; **n:** Billion roubles; **p:** et material product; **q:** In roubles; **s:** Currency: Ugandan shillings; Currency: Roubles.

Byelorussian Soviet Socialist Republic (r) / Ukrainian Soviet Socialist Republic (r)

Byelorussian SSR (r)			Ukrainian SSR (r)		
1976	1981	1986	1976	1981	1986
9.39	9.68	10.08	49.1	50.1	51.0
...	...	...	...	...	...
...	...	...	...	...	...
...	...	...	47.6	48.8	48.4[b]
...	...	...	26.2	22.4	21.3[b]
...	...	...	37.5	38.9	39.4[b]
51	57	64	59	63	67
...	...	...	0.2	0.4	0.4
...	66/76[d]	67/76[d]	67/74	62/73	64/74
...	...	...	...	16.6[a]	14.8
...	...	...	69.8[pq]	81.4[pq]	97.7[p
...	...	...	...	3.1[p]	3.7[p
...	...	...	1 422[q]	1 649[q]	1 955[q
...	...	...	27.5	22.9	27.0
...	...	...	...	...	...
...	...	...	...	...	...
...	...	...	...	...	...
...	...	...	95.3	100	113[b
...	...	...	...	...	...
...	...	...	86	103	118[b
...	...	...	...	119[hk]	140[
...	...	...	...	...	...
...	...	...	...	...	...
96.7	101.2	107.1	96.5	101.0	105.4
...	...	...	...	...	...
0.748[e]	0.720[e]	0.684[e]	0.748[e]	0.720[e]	0.684[e
...	...	...	...	...	...
...	...	...	...	...	...
...	...	...	...	...	...
...	...	...	...	...	...
...	...	...	...	...	...
188[c]	228	249[b]	234	254[a]	295[
...	...	...	...	...	...
...	...	...	...	...	...
...	...	...	...	3.1	...
332[c]	289	266[b]	321[c]	378	420[
...	...	...	...	...	...

a: 1980; b: 1985; c: 1975; e: Currency: Roubles; h: Ba
1980 = 100; k: Index numbers of electric energy production; p: Net material produ
q: In roubles; r: Data included with USSR.

nited Arab mirates

United Kingdom

1976	1981	1986	1976	1981	1986
0.59a	1.06a	1.38a	56.21	56.35	56.76
28.2	28.6	31.8	23.3	20.9	19.5
2.9/4.7	1.7/2.8	2.1/3.0	16.6/22.5	17.0/23.1	17.6/23.7
74/6	74/9b	...	59/34c	60/36	59/37d
4.6	4.6b	...	2.5	2.1	2.3d
10.2	10.3b	...	31.1	27.0	24.1d
79.8	81.2	77.8	89.8	90.7	91.5
13.3	6.1	3.5	-0.0	0.1	0.0
64/68	65/70	67/72	70/76	71/77	71/78
46	38	32	14	10	9
9 962c	29 629b	27 081e	235 338c	534 395b	454 540e
63.8	14.4	-2.2	2.5	1.9	2.0
19 727c	30 234b	20 408e	4 186c	9 518b	8 069e
32	25	29d	19	16	17e
1	1	1d	2	2	2e
65	66	58d	31	31	32e
1	7	10d	26	22	23e
...	...	...	82	101	112
...	...	...	82	101	112
...	...	...	95	96	110
96 069	81 988	74 540	121 511	198 335	250 257
3 327	9 646	6 791e	55 755	102 725	126 208
8 684	20 240	13 124e	45 372	102 820	107 013
...	...	...	59.6	111.9	146.3
...	...	...	...	111f	127f
3.977p	3.671p	3.671p	0.587q	0.524q	0.678q
1 907	3 202	3 370	3 375	15 238	18 422
0.54	0.68	0.82	21.03	19.03	19.01
...	...	...	10 071g	11 453g	13 772g
63c	253	...	250c	313	348e
73c	231	208e	364c	497	524d
50c	95b	98e	360c	404b	437e
42/62c	...	...	...	...	...
73/70c	79/76b	79/86e	93/94c	92/93b	91/93d
0.8c	1.1b	1.8e	...	12.0	11.6d
950	878h	...	677k	626n	...
3 565	3 594	3 644	3 244	3 174	3 130

United Nations estimate; **b:** 1980; **c:** 1975; **d:** 1984; **e:** 1985; **f:** Agricultural pro-
cts; **g:** Departures; **h:** 1982; **k:** 1977; **n:** 1979; **p:** Currency: Dirhams; **q:** Currency:
unds sterling.

United Republic of Tanzania / United States

United Republic of Tanzania			United States		
1976	1981	1986	1976	1981	1986
16.41	19.17	22.46	218.03	230.04	241.60
47.9	48.4	48.8	25.2	22.5	21.9
3.4/4.2	3.4/4.2	3.4/4.1	12.9/16.7	13.6/17.8	14.0/18.4
...	...	...	57/36[a]	57/40	57/42
...	...	...	3.7	2.7[b]	3.1
...	...	...	24.5	23.0[b]	21.4
10.1	16.5	22.3	73.7	73.7	73.9
3.4	3.5	3.6	1.1	0.9	0.9
47/51	49/53	51/55	69/77	71/78	71/79
125	115	106	14	11	10
2 692[g]	5 138[b]	6 401[c]	1 583 920[g]	2 658 470[b]	3 959 610
4.3	2.9	0.4	2.5	3.5	3.0
169[g]	272[b]	285[c]	7 334[g]	11 805[b]	16 636
21	20	11[c]	17	19	19
37	41	53[c]	3	3	2
13	11	6[c]	29	29	27
12	9	5[c]	23	22	20
89	103	111	92	106	102
100	99	92	96	104	98
...	...	...	85	102	115
44	50	56	1 346 817	1 397 626	1 395 378
567	1 212	780	129 896[d]	273 352[d]	387 054
459	613	346	115 413[d]	233 739[d]	217 336
54.3	125.6	280.2[e]	69.1	110.4	133.1
...	...	...	...	109[f]	111
8.324[p]	8.322[p]	51.717[p]	1.00[q]	1.00[q]	1.00
112	19	61	7 149	18 924	37 452
...	...	...	274.68	264.11	262.04
98	92	59[c]	17 523	23 080	25 359
3[g]	5	...	491[g]	693	...
4[g]	5	5[c]	686[g]	789	...
0.3[g]	0.4[b]	0.5[c]	560[g]	684[b]	798
38/69[k]	...	...	...	...	...
39/27[g]	62/53[b]	45[ch]	97/96[g]	95/96[b]	101/100
4.4[g]	4.3[b]	...	5.8[g]	...	...
24 029	...	...	610[g]	549	...
2 257	2 427	2 314	3 305	3 606	3 652

a: 1977; **b:** 1980; **c:** 1985; **d:** Including trade of U.S. Virgin Islands and Puerto R but excluding trade between USA and its other dependencies/Beginning 1977 incluc non-monetary gold; **e:** 1984; **f:** Domestic supply; including exported goods; **g:** 19 **h:** Both sexes; **k:** 1978; **p:** Currency: Tanzanian shillings; **q:** Currency: United Sta dollars.

ruguay			Vanuatu		
1976	1981	1986	1976	1981	1986
2.85	2.93	2.98	0.10	0.12[a]	0.14
27.7	27.1	26.9	...	...	...
.9/15.4	13.2/16.0	13.8/16.9	...	...	...
57/22[a]	...	54/32[b]	...	49/42[c]	...
16.0[a]	...	2.8[b]	...	76.8[c]	...
20.5[a]	...	21.8[b]	...	2.2[c]	...
83.0	83.8	84.6	...	...	...
0.6	0.7	0.8	...	...	...
66/73	67/74	68/74	...	...	...
44	30	27	...	...	...
3 623[a]	10 133[p]	5 054[d]	...	...	...
1.3	4.4	-3.9	...	...	...
1 281[a]	3 485[p]	1 678[d]	...	...	...
15	16	8[d]	...	17	12[b]
10	8	10[d]	...	21	18[b]
23	22	28[d]	...	14	12[b]
21	20	24[d]	...	4	3[b]
107	114	107	89	107	111
112	114	98	104	103	89
82	96	86	...	...	...
103	215	627	...	...	...
587	1 641	820	30	58	57
546	1 215	1 087	15	32	17
16[f]	134[f]	1 123[f]	60.9[gh]	111.1[gh]	121.1[gh]
...	123[k]	1 266[k]	...	...	...
4.0[ns]	11.594[ns]	181.00[ns]	...	91.230[t]	110.320[t]
176	430	482	...	8	21
3.54	3.39	2.60	...	...	...
492	928	1 168	...	22	18
...	92	117[d]	34[a]	46	25[d]
90[a]	101	128[d]	20[a]	...	24
124[a]	125[p]	166[d]	...	...	...
7/6[a]	...	...	...	43/52	...
84[aq]	84/84[p]	90[bq]	...	...	...
...	2.3	2.5[d]	...	...	...
710	536[c]	...	...	4 379[r]	...
2 944	2 832	2 721	2 306	2 403	2 331

1975; **b:** 1984; **c:** 1979; **d:** 1985; **e:** Manufacturing; **f:** Montevideo; **g:** Base: 81 = 100; **h:** Urban areas, low income group; **k:** Montevideo; Domestic supply; in- ding production of exports; excluding mining and quarrying; **n:** Selling rate; **p:** 1980; Both sexes; **r:** 1982; **s:** Currency: New Pesos; **t:** Currency: Vatu.

Venezuela

1976	1981	1986
13.12	15.48	17.79
43.5	41.1	39.5
4.5/5.2	4.6/5.4	4.9/5.7
45/17	47/18	49/19c
18.6b	14.2	14.5c
17.8b	18.3	18.3c
80.2	83.7	86.6
3.4	2.8	2.6
65/71	66/72	67/73
43	39	36
27 561b	59 213n	49 604c
5.0	3.3	-1.6
2 176b	3 941n	2 864c
32	24	15c
6	6	8c
41	40	39c
17	15	21c
87	99	111
99	97	93
57d	110d	176cd
135 448	130 314	119 111
6 023e	11 813e	8 600e
9 570	20 125	10 052
63.4	116.0g	124.3fg
...	114h	211h
4.29ks	4.29ks	14.50ks
8 124	8 164	6 437
11.18	11.46	11.46
535	200	317
75b	148	195c
54b	56	83c
101b	114n	130c
...	13/17	...
75/79	76/81q	77/81c
5.0b	5.0n	6.0p
930	947r	...
2 444	2 665	2 550

Viet Nam

1976	1981	1986
48.80a	55.31a	60.92a
43.7	42.5	39.3
5.8/7.0	5.6/6.7	5.6/6.7
54/40b	...	...
...	...	...
...	...	...
18.8	19.3	20.3
2.4	2.0	2.0
54/58	57/61	59/63
90	76	67
...	...	...
...	...	...
...	...	...
...	...	...
...	...	...
...	...	...
...	...	...
81	105	135
89	103	119
...	...	...
3 899	4 262	4 022
728b	996	...
107b	159	...
...	...	...
...	...	...
1.834t	9.045t	15.00
...	...	...
...	...	...
...	...	...
...	...	...
...	...	2
...	...	33
...	9/22	...
71/73b	72/67n	72/65
...	...	...
5 668b	4 068	...
...	...	...

a: United Nations estimate; **b:** 1975; **c:** 1985; **d:** Manufacturing; **e:** F.O.B.; **f:** Ba 1984 = 100; **g:** Caracas; metropolitan area; **h:** Domestic supply; excluding produ of mining and quarrying; **k:** Selling rate; **n:** 1980; **p:** 1984; **q:** 1982; **r:** 1979; **s:** Curr cy: Bolívares; **t:** Currency: Dong.

Yemen

1976	1981	1986
5.38[a]	6.14[a]	7.05[a]
46.4	46.8	46.9
5.4/5.4	5.5/5.5	5.5/5.5
46/6[b]	...	...
73.6[b]	...	...
3.2[b]	...	...
11.0	15.3	20.0
2.5	2.7	2.9
45/47	47/50	49/52
150	135	120
1 082[b]	2 768[d]	...
9.8	6.1	...
205[b]	463[d]	...
23	42	...
36	28	...
6	9	...
5	7	...
92	106	139
101	103	119
...	...	...
...	...	...
413	1 758	1 289[c]
8	47	10[c]
...	...	...
...	...	...
4.56[h]	4.56[h]	8.10[ch]
720	962	432
0.00	0.01	0.00
...	56	44
5[b]	11	...
...	...	...
-	0.8[d]	4.1[c]
...	...	...
34/4[b]	54/9[d]	...
1.1	...	...
20 950[b]	11 670	...
2 035	2 197	2 254

Yugoslavia

1976	1981	1986
21.57	22.47	23.27
25.7	24.8	23.8
11.2/14.0	10.0/13.1	10.6/14.8
56/31	54/33	...
...	28.7	...
...	23.6	...
38.5	42.3	46.3
0.9	0.8	0.6
68/73	68/73	69/75
35	30	25
33 280[b]	69 958[d]	44 238[c]
5.9	6.1	0.6
1 559[b]	3 137[d]	1 911[c]
30	28	21[c]
14	13	11[c]
31	34	39[c]
26	29	34[c]
96	101	103
99	100	98
75	104	119
17 782	23 444	24 715
7 367	15 817	8 315
4 878	10 929	7 206
48	141	1 312
...	...	...
18.231[k]	41.823[k]	457.180[k]
1 990	1 597	1 460
1.47	1.86	1.87
5 573	6 616	8 464
72[b]	125	134[c]
60[b]	102	132[f]
144[be]	...	175[ce]
...	...	...
87/82[b]	90/87[g]	88/85[c]
4.8[b]	4.0[d]	3.4[c]
810[b]	672	...
3 496	3 587	3 599

United Nations estimate; **b:** 1975; **c:** 1985; **d:** 1980; **e:** Number of licences issued
sets declared; **f:** 1984; **g:** 1982; **h:** Currency: Yemeni rials; **k:** Currency: New dinars.

Zaire 1976	Zaire 1981	Zaire 1986	Zambia 1976	Zambia 1981	Zambia 1986
23.29	26.61a	30.85a	5.14	5.83a	6.90a
44.6	45.0	45.2	46.5	46.9	47.3
3.9/5.3	4.0/5.3	4.0/5.3	3.9/4.6	4.0/4.6	4.0/4.7
51/37	...	...	44/17b	45/17	...
...	...	...	...	...	...
...	...	...	...	...	...
32.2	34.2	36.6	36.3	42.8	49.5
2.9	2.9	3.0	3.1	3.3	3.4
46/50	48/52	50/54	48/51	50/53	52/55
117	107	98	94	88	80
3 851n	...	...	2 463n	3 883k	2 597p
3.1	...	...	2.9	-0.6	-0.2
172n	...	...	509n	688k	390p
23	...	...	24n	18k	10p
26	...	...	15	16	14p
20	...	...	36	36	37p
8	...	...	15	20	22p
97	103	118	120	100	119
108	100	98	137	97	96
129c	102c	...	110	98	88
1 665	1 497	1 764	1 038	1 121	1 198
674	672	884	655r	1 062r	582r
930	662	1 092	1 040	1 074	434
13.5e	135.4e	...	58.6f	114.0f	383.6f
...	...	...	...	112g	557g
0.861v	5.465v	71.100v	0.793x	0.883x	12.710x
50	152	269	93	56	70
0.26	0.36	0.47	0.17	0.22	0.00
19	24	31	56h	147h	100h
...	6	4q	18n	28	...
2n	1k	1p	11n	10	10p
0.4n	0.4k	0.4p	4.6n	11k	14p
...	...	...	...	...	...
75/47n	86/54k	...	74/60n	73/63k	72d
3.1n	3.4k	...	5.2n	4.2k	5.1q
...	13 453t	...	...	7 106	...
2 292	2 127	2 154	2 320	2 203	2 123

a: United Nations estimate; **b:** 1977; **c:** Manufacturing; **d:** Both sexes; **e:** Kinshas
f: Low income group; **g:** Domestic supply; **h:** Visitor arrivals; **k:** 1980; **n:** 1975; **p:** 198
q: 1984; **r:** F.O.B.; **t:** 1979; **v:** Currency: Zaire; **x:** Currency: Kwacha.

Zimbabwe

1976	1981	1986
6.33	7.36	8.41
46.9	47.3	47.6
4.1/4.7	4.1/4.7	4.0/4.7
...	...	...
...	...	...
...	...	...
19.4	21.9	24.6
3.4	3.5	3.6
52/56	54/58	56/60
86	80	72
3 518[d]	5 351[c]	5 024[e]
4.6	0.1	3.9
566[d]	726[c]	572[e]
20	19	17[e]
16	15	12[e]
32	30	36[e]
22	23	26[e]
104	111	123
124	118	99
92	106	112
2 882	1 973	3 104
620[a]	1 472[a]	985[a]
900	1 406	1 019
68.9[b]	113.1[b]	229.7[b]
...	...	...
0.619[h]	0.717[h]	1.678[h]
77	169	106
0.26	0.47	0.54
...	325	395
...	33	34[f]
28[d]	29	32[e]
9.2[d]	10[c]	14[e]
...	...	...
52/43[d]	96/86[g]	102/91
3.3[d]	6.4[c]	...
...	6 411[c]	...
2 118	2 111	2 094

F.O.B.; **b:** Low-income group; **c:** 1980; **d:** 1975; **e:** 1985; **f:** 1984; **g:** 1983; **h:** Cur-
ncy: Zimbabwean dollars.

General notes

Geographical coverage

The designations employed and presentation of the material in this publication were adopted solely for the purpose of providing a convenient geographical basis for the accompanying statistical series. The same qualification applies to all notes and explanations concerning the geographical units for which data are presented.

Because of space limitations, the country and region names listed in the tables are generally the commonly employed short titles in use in the United Nations, the full titles being used only when a short form is not available.
Countries or regions are listed in alphabetical order. The regions are shown in the following order: Africa; America, North; America, South; Asia, excluding the USSR; Europe, excluding the USSR; and Oceania.

"America, North" comprises Canada, the United States, Mexico and the countries in Central America and the Caribbean. "America, South" covers the rest of America. "America, Northern" comprises Canada and the United States. "America, Latin" comprises "America, South", Mexico and countries in Central America and the Caribbean. "East Asia" comprises China, Japan, Hong Kong, Korea (Democratic People's Republic of Korea and Republic of Korea), Macau and Mongolia. "South Asia" covers the rest of Asia. "Europe" comprises all of Europe except the USSR.

Data relating to the People's Republic of China generally include those for Taiwan Province, in the fields of statistics relating to population, area, natural resources, natural conditions such as climate etc. In other fields of statistics, they do not include Taiwan Province unless specifically stated.

The data which relate to the Federal Republic of Germany and the German Democratic Republic include the relevant data relating to Berlin for which separate data have not been supplied. This is without prejudice to any question of status which may be involved.

The data shown for the Byelorussian Soviet Socialist Republic and the Ukrainian Soviet Socialist Republic are also included in the data for the Union of Soviet Socialist Republics.

Technical notes

Area

Surface area data represent the total surface area, comprising land area and inland waters (assumed to consist of major rivers and lakes) and excluding only polar regions and uninhabited islands.

Agricultural area includes arable land, land under permanent crops and permanent meadows and pastures, but excludes land under trees, grown for wood and timber.

Forests and woodlands refer to land under natural or planted stands of trees, whether or not productive, and include land from which forests have been cleared but that will be reforested in the foreseeable future.

Demography and labour force

Total population estimates are generally for the mid-year, de facto (present-in-area) population. Official country or area estimates are supplemented for some countries or areas by the United Nations estimates. These are foot-noted.

Ratio of children population (aged 0-14 years) refers to total population without regard to sex; elderly population (60 years and over) refers to elderly males as a percentage of all males and elderly females as a percentage of all females.

Density data refer to number of persons in total population per square kilometre of total surface area. Density figures are not to be considered as either reflecting density in the urban sense or as indicating the supporting power of a territory's land and resources.

The economically active population is the total of employed persons (including employers, persons working on their own account, salaried employees and wage earners and, in so far as data are available, unpaid family workers) and of unemployed persons at the time of the census or survey. In general, the economically active population does not include students, women occupied solely in domestic duties, retired persons living entirely on their own means and persons wholely dependent upon others. Percentages generally refer to both males and females. If distribution by sex is not available, data refer to totals only.

Economically active population in agriculture refers to population in agriculture, hunting, forestry and fishing; in the case of industry, it includes mining, quarrying, manufacturing, electricity, gas and water.

The comparability of labour force data is hampered by differences between countries -- and even within a country -- not only different details of definitions used and groups covered, but also differences in the methods of collection, classification and tabulation of data. For some countries, statistics cannot be compared over time since they are drawn from a variety of sources (ILO estimates, census data, survey data, official national estimates) -- all of which differ as to scope, coverage, reference period, time of year etc. In sum, the differences in the levels shown may in large measure be due to statistical practices rather than real changes. Readers should treat some labour force data with due caution because of these differences in sources and national practice.

The *urban population* percentages (rates) are based on the number of persons defined as "urban" according to national definitions of this concept. In most cases these definitions are those used for the most recent population census.

The *annual population growth rate* is the average annual percentage change in population size between the designated years. Although the population estimates appear only in millions to decimal places, figures in units were used for the computation whenever possible.

The *crude birth and death rates* indicate the number of live births and deaths in a year per thousand mid-year population. The *infant mortality* rate is the number of infants who die before reaching 1 year of age, per thousand live births in a given year or period. The *life expectancy at birth* indicates the average number of years expected to be lived by a newly-born baby, assuming a fixed schedule of age-specific mortality rates.

National accounts

The estimates of total and per capita gross domestic product (GDP) in purchasers' values are expressed in national currency. They should be considered as indicators of the total and per capita production of goods and services of the countries represented and not as measures of the standard of living of their inhabitants.

Gross domestic product in purchasers' values is defined as (a) the difference between the gross output value of resident producers minus the value of their intermediate consumption plus import duties, (b) the sum of domestic factor incomes (i.e., compensation of employees and operating surplus), indirect taxes net of subsidies and consumption of fixed capital, (c) the sum of government and private final consumption expenditure, gross capital formation in fixed assets and stocks and exports minus imports of goods and services.

For the countries with centrally planned economies, data in general are shown in terms of the System of Material Product Balances (MPS). Net material product (NMP) is defined as (a) the global product (gross output) of goods and material services less the intermediate consumption, less the consumption of fixed assets, (b) the sum of primary incomes of the population and enterprises, (c) the final uses of goods and material services--final consumption, net capital formation, losses and the excess of exports over imports. Net material product data are shown in national currencies.

The *annual growth rates of GDP and NMP* are computed as average annual geometric rates of growth expressed in percentage form for the periods indicated. They are based on the estimates of GDP and NMP at constant prices.

Gross (or net) fixed capital formation, agriculture, total industrial activity (mining and quarrying, manufacturing, electricity, gas and water) and *manufacturing* data are the percentage distribution of GDP or NMP in current prices.

Agriculture and industry

The *index numbers of agricultural production* cover all crops and livestock products originating in each country on which information is available. Indexes for regions and the world are computed by summing the country aggregates, which for that purpose are converted into US dollars.

The index numbers of food production cover commodities that are considered edible and contain nutrients. Accordingly, coffee and tea are excluded because, although edible, they have practically no nutritional value. The index numbers shown may differ from those produced by countries themselves because of differences in concepts of production, coverage,weights, time reference of data, and methods of evaluation.

The index numbers of industrial production generally cover mining, manufacturing and electricity, gas and water, and do not include construction unless otherwise indicated.

Commercial energy production refer to the production of various forms of primary energy which is converted into a common unit (metric ton of oil equivalent).

Agricultural production statistics are presented on a calendar-year time-reference basis. That is to say, the data for each particular crop refer to the calendar year in which the entire or the bulk of the harvest of the crop takes place.

Data on *livestock numbers* cover selected domestic animals irrespective of their age and place and/or purpose of their breeding. Estimates have been made for non-reporting countries as well as for countries reporting partial coverage.

Fish catches data cover, as far as possible, live weight of catches in both sea and inland fisheries. They generally include seaweed, crustaceans and molluscs, but not aquatic mammals (whales, dolphins etc.).

Data for the *production of industrial commodities* refer to the total industrial production of each commodity during the years indicated. Data include the production of the industrial establishments in which the commodity is a primary, secondary, or intermediate product. Unless otherwise indicated, production by non-industrial establishments is excluded.

Trade, finance and tourism

External trade is the movement of goods into (imports) and out of (export) a country. Exports are generally valued at the frontier of the exporting country (f.o.b. valuation). Imports are valued at the frontier of the importing country (c.i.f. valuation). Both imports and exports are shown in US dollars. Conversion from national currencies is made by means of currency conversion factors based on official exchange rates (par values or weighted average exchange rates).

Consumer price index numbers are designed to show changes over time in the cost of selected goods and services that are considered as representative of the consumption habits of the population concerned. The indexes generally refer to "all items" and to the country as a whole. When the index refers to a city, the name of the city is indicated.

Wholesale price index numbers are designed to measure changes in the level of commodity prices at a non-retail stage of distribution. Data usually represent a mixture of producer and wholesale prices for domestic goods. In addition, these indexes may cover the prices of goods imported in quantity for the domestic market. Unless otherwise stated data usually refer to the country as a whole.

Exchange rates are the amount of one currency required to purchase a fixed amount of another currency. Data are generally shown in units of national currency per one US dollar and refer to the end-of-period quotation. For most countries, mid-point rates (i.e., the average of buying and selling rates) are shown.

International reserves minus gold are the sum of the US dollar value of monetary authorities, holdings of special drawing rights (SDRs), reserve position in the International Monetary Fund and the foreign exchange.

Gold reserves series report data on official holdings of gold expressed in physical terms of fine troy ounces.

Tourists are persons travelling for pleasure, domestic reasons, health, meetings, business, study etc., and stopping for a period of 24 hours or more in a country or area other than that in which they usually reside. They do not, therefore, include immigrants, residents in a frontier zone, persons domiciled in one country or area and working in an adjoining country or area, travellers passing through a country or area without stopping, transport crews and troops.

Transport

The *railways-freight traffic* figures relate to the domestic and international traffic on all railway lines within the countries irrespective of gauge, except railways entirely within an urban unit; railways serving a plantation, forest, mine or industrial plant; and funicular and cable railways.

The *international sea-borne shipping* figures relate to goods loaded and unloaded by dry cargo, crude petroleum and petroleum products. Petroleum products exclude bunkers and those products not generally carried by tanker. Data represent the weight of all goods (including packing) and livestock in external trade loaded and unloaded from sea-going vessels of all the ports of the countries.

The *civil aviation series* refer to the scheduled traffic of the airline or airlines registered in countries for total operation (international and domestic) irrespective of where the movement of persons and goods takes place.

The *motor vehicles in use* series refer to passenger cars and commercial vehicles in use according to census or registration figures for years census or annual registration took place. Passenger vehicles are vehicles seating no more than nine persons (including the driver) including taxis, jeeps and station wagons. Commercial vehicles include vans, lorries (trucks), buses, tractors and semi-trailer combinations. In country chapters, the data are presented for motor vehicles in use per thousand inhabitants.

Consumption (Per caput)

Sugar: The series relate to the consumption of centrifugal sugar, including sugar used for the manufacture of sugar-containing products and sugar for purpose other than human consumption as food. In general, data are expressed in terms of raw value.

Fertilizers: The series relate to the use of total fertilizers in agriculture, thus excluding materials used for technical purposes. Data refer to fertilizer year 1 July - 30 June.

Steel: "Apparent consumption" expressed in terms of crude steel equivalent.

Newsprint: Data represent "apparent consumption".

Commercial energy: Data are based on the consumption of solid, liquid and gaseous fuels as well as electricity, presented in oil equivalent.

Education, health and nutrition

Adult illiteracy ratio: Data refer to percentage of illiterate adult population (aged 15 and over) for both sexes. Ability to both read and write is used as a criterion of literacy, hence, all semi-literates--persons who can read but not write--are included with illiterates.

School enrolment ratios (gross): For the first and second levels, the enrolment ratio generally is the total enrolment of all ages in first- and second-level education, divided by the total population in the specific age group used by the country. For the third-level, the enrolment ratio is the total enrolment of all ages in the third-level education divided by the total population in the 20-24 year age group. In some cases, the gross enrolment ratio will exceed 100 if the actual age distribution of pupils spreads over outside the official school ages.

Public expenditure on education: This series shows the general trends in public expenditure on public and private education expressed as a percentage of the gross national product. The data shown should be considered as approximate indications of the public resources allocated to education. These data are revised every year and may sometimes differ from those shown in previous editions of the *Pocketbook*.

Physicians are persons who are graduates of a medical school or faculty actually working in the country in any medical field (practice, teaching, administration, research, laboratory etc.).

Food supply: Data provide the daily per capita food supply available for human consumption in calories.

Communication

Radio receivers in use: All types of receivers for radio broadcasts to the general public, including those connected to a cable distribution sytem (wired receivers) per thousand inhabitants.

Television receivers in use: Television receivers in use and (or) licences issued per thousand inhabitants.

Telephones in use: The number of public and private telephones installed which can be connected to a central exchange per thousand inhabitants.

Conversion coefficients and factors

The metric system of weights and measures has been employed in the Pocketbook. The following table shows the equivalents of the basic metric, British imperial and United States units of measurement:

Area		
1 square km.	=	0.386 102 square mile
1 hectare	=	2.471 054 acres
Volume		
1 cubic metre	=	35.314 667 cubic feet or
	=	1.307 951 cubic yards
Weight or mass		
1 ton	=	1.102 311 short tons, or
	=	0.984 207 long tons
1 gram	=	0.032 951 troy ounce
1 kilogram	=	35.273 962 avdp. ounces
	=	2.204 623 avdp. pounds
Railway and air transport		
1 kilometre	=	0.621 371 mile
1 passenger-Km.	=	0.621.371 passenger-mile
1 ton-kilometre	=	0.684 944 short ton-mile, or
	=	0.611 588 long ton-mile

Symbols, conventions and abbreviations

.. or ...	Data not available
-	Magnitude zero or negligible or data not applicable
0 or 0.0	Magnitude not zero, but less than half of the unit employed
Per cent: M/F	Per cent: Males/Females
%	Per cent
Per cent pa	Per cent per annum
Km2	Square kilometre
HA	Hectare
1 000 MT	Thousand metric tons
Mill MT	Million metric tons
Mill Cu M	Million cubic metres
1 000 TJ	Thousand terajoules
Billion Kwh	Billion kilowatt hours

Decimal figures are always preceded by a period (.). Thousands and millions are separated by a space.

Selected list of recurrent publications

Monthly Bulletin of Statistics

(Series Q) (E/F)
Provides monthly statistics on 69 subjects from over 200 countries and territories, together with special tables illustrating important economic developments. Quarterly data for significant world and regional aggregates are included regularly.

Statistical Yearbook

(Series S) (E/F)
A comprehensive compilation of international statistics relating to: population and manpower; national accounts; wages, prices and consumption; government accounts; balance of payments; finance; health; education; culture; science and technology; housing; development assistance; industrial property; agriculture, forestry and fishing; industrial production; energy; construction; wholesale and retail trade; external trade; communications; transport and international tourism.

World Statistics in Brief (United Nations Statistical Pocketbook),

(Series V) (E)
A series of annual compilations of basic international statistics. Part of the Pocketbook presents information on each of 167 countries, showing important and frequently consulted statistical indicators. Another part contains demographic, economic and social statistics for the world as a whole and for continents.

Demographic Yearbook

(Series R) (E/F)
A comprehensive compilation of international demographic and related population statistics for about 220 countries or areas of the world. It contains tables on population, natality, mortality, nuptiality and divorce as well as tables on special topics.

Population and Vital Statistics Report

(Series A) (E) Issued quarterly
This publication presents, in one general table, series on total population and on births, deaths and infant deaths for over 200 countries or areas of the world.

E = English; F = French; R = Russian; S = Spanish; E/F = Bilingual text.

National Accounts Statistics (series)

National Accounts Statistics: Main Aggregates and Detailed Tables

(Series X) (E)

This contains detailed national accounts estimates for 156 countries and areas. The data for each country or area are presented in separate chapters under uniform table headings and classifications recommended in the System of National Accounts (SNA) for countries with market economies or the System of Material Product Balances (MPS) for countries with centrally planned economies.

National Accounts Statistics: Analysis of Main Aggregates

(Series X) (E)

This contains a summary of main national accounts aggregates in the form of twelve analytical tables. Data are analyzed for gross domestic product or net material product by type of expenditure, kind of economic activity and cost components and other principal income aggregates, based on current prices and constant prices.

National Accounts Statistics: Government Accounts and Tables

(Series X.3) (E)

This publication is devoted to the government sector and aims to assemble a complete set of accounts and tables for the general government and its subsectors within the system of national accounts in order to provide a comprehensive profile of the sector and to facilitate analysis of its activities.

International Trade Statistics Yearbook

(E/F) Volumes I and II.

Volume I. Trade by Country

(Series G) (E/F)

This publication provides the basic information for individual countries' external trade performance in terms of overall trends in current value as well as in volume and price, the importance of trading partners and the significance of individual commodities imported or exported. Also published are basic summary tables showing, *inter alia,* the contribution of trade of each country to the trade of its region and of the world, analysing the flow of trade between countries and describing the fluctuations of the prices at which goods moved internationally.

Volume II. Trade by Commodity: Commodity Matrix Tables

(Series G) (E/F)

This contains a chapter on price and price indexes, commodity tables showing the total economic world trade of certain commodities, analysed by regions and countries, and commodity matrix tables.

ommodity Trade Statistics

eries D) (E)

Issued in fascicles of about 250 pages as annual data become available in the 625 sub-groups of the United Nations Standard International Trade Classification (SITC). Data are published in SITC Revision 2 as they become available. Approximately 28 fascicles annually.

Quarterly and annual data are available on microfiches.

ergy Statistics Yearbook

eries J) (E/F)

This is an internationally comparable series on commercial energy, summarising world energy trends. Annual data are presented on the production, trade and consumption of solid, liquid and gaseous fuels and electricity. Global and regional totals and per capita consumption series are also included, as are data on energy reserves and energy prices.

ergy Balances and Electricity Profiles

eries W,) (E)

This is a compilation of energy data for developing countries in the format of overall energy balances and electricity profiles.

dustrial Statistics Yearbook

Volume I. General Industrial Statistics

(Series P) (E)

This is a series of annual compilations on world industry. It presents information on major items of industrial activity classified by branches of activity. Also included are international tables on index numbers of industrial production and employment.

Volume II. Commodity Production Statistics

(Series P) (E)

This presents internationally comparable information on the production of about 534 industrial commodities (575 statistical series) for approximately 200 countries and regions. The commodities have been selected on the basis of their importance in world production and trade and represent the principal products of mining and manufacturing.

nstruction Statistics Yearbook

eries U) (E)

Annual compilations on construction statistics. It contains data for about 135 countries on general indicators of construction activity, fixed assets, building and dwelling construction authorized and completed and index numbers of construction activity.

40257—OCTOBER 1988—7,500M